REPRESENTATION RECO[illegible]

Cognitive representation is the single most important explanatory notion in the sciences of the mind and has served as the corner-stone for the so-called "cognitive revolution." This book critically examines the ways in which philosophers and cognitive scientists appeal to representations in their theories, and argues that there is considerable confusion about the nature of representational states. This has led to an excessive over-application of the notion – especially in many of the newer theories in computational neuroscience. *Representation Reconsidered* shows how psychological research is actually moving in a non-representational direction, revealing a radical, though largely unnoticed, shift in our basic understanding of how the mind works.

WILLIAM M. RAMSEY is Associate Professor in the Department of Philosophy, University of Notre Dame.

REPRESENTATION RECONSIDERED

WILLIAM M. RAMSEY

University of Notre Dame

CAMBRIDGE UNIVERSITY PRESS
Cambridge, New York, Melbourne, Madrid, Cape Town, Singapore,
São Paulo, Delhi, Dubai, Tokyo, Mexico City

Cambridge University Press
The Edinburgh Building, Cambridge CB2 8RU, UK

Published in the United States of America by Cambridge University Press, New York

www.cambridge.org
Information on this title: www.cambridge.org/9780521153324

First published 2007
First paperback printing 2010

A catalogue record for this publication is available from the British Library

ISBN 978-0-521-85987-5 Hardback
ISBN 978-0-521-15332-4 Paperback

For my parents,
Jim and Diane Ramsey
and for Mimi

Contents

Figures

Preface

It has become almost a cliché to say that the most important explanatory posit today in cognitive research is the concept of representation. Like most clichés, it also happens to be true. Since the collapse of behaviorism in the 1950s, there has been no single theoretical construct that has played such a central role in the scientific disciplines of cognitive psychology, social psychology, linguistics, artificial intelligence, and the cognitive neurosciences. Of course, there have been many different types of representational theories. But all share the core assumption that mental processes involve content-bearing internal states and that a correct accounting of those processes must invoke structures that serve to stand for something else. The notion of mental representation is *the* corner-stone of what often gets referred to in Kuhnian terms as the "cognitive revolution" in psychology. But mental representation hasn't been important just to psychologists. Accompanying this trend in the sciences has been a corresponding focus on mental representation in the philosophy of mind. Much of this attention has focused upon the nature of commonsense notions of mental representation, like belief and desire, and how these can be part of a physical brain. More specifically, the central question has focused on the representational nature of beliefs – the fact that they have meaning and are essentially *about* various states of affairs.

Yet despite all of this attention (or perhaps because of it), there is nothing even remotely like a consensus on the nature of mental representation. Quite the contrary, the current state of affairs is perhaps best described as one of disarray and uncertainty. There are disagreements about how we should think about mental representation, about why representations are important for psychological and neurological processes, about what they are supposed to do in a physical system, about how they get their intentional content, and even about whether or not they actually exist. Part of this chaos is due to recent theoretical trends in cognitive science. The central explanatory framework behind a great deal of cognitive research has traditionally been the classical computational theory of cognition. This framework regards the mind as a computational system with discrete internal symbols serving as representational states. However, over the past twenty years there have been dramatic departures from the classical computational framework, particularly with the emergence of theories in the cognitive neurosciences and connectionist modeling. These newer approaches to cognitive theorizing invoke radically different notions of cognitive representation; hence, they have generated considerable disagreement about how representation should be understood.

Still, debates over representation are not simply due to the existence of different cognitive theories and models. Often, the nature of representation *within* these different frameworks is unclear and disputed. One might expect some assistance on these matters from philosophers of psychology, especially given the amount of philosophical work recently focusing upon representation. Yet up to this point, it is far from obvious that philosophical work on representation has helped to ameliorate the situation in cognitive science. Philosophical work on representation has been a predominantly *a priori* enterprise, where intuitions about meaning are analyzed without special concern for the nuances of the different notions of representation that appear in scientific theories. While abstract questions about the nature of content are important, esoteric discussions about hypothetical scenarios, like the beliefs of Twin-Earthlings or spontaneously generated "swamp-men," have failed to be of much use to non-philosophers in the scientific community. Moreover, because of a preoccupation with the nature of content, philosophers have neglected other issues associated with cognitive representation that are more pressing to researchers. Of these other issues, perhaps the most important is explaining what it is for a neurological (or computational) state actually to *function as a representation* in a biological or computational system. Despite the importance of this issue to empirical investigators, the actual role representations are supposed to play, *qua representations*, is something that has received insufficient attention from philosophers.

My own interest in these matters began as a graduate student in the mid-1980s, with a front row seat on the exciting development of connectionist modeling taking place at the University of California, San Diego. A great deal of buzz was generated by the radically different picture of representation that accompanied connectionist models, especially their distributed and non-linguistic form. Yet every time I tried to get a clearer sense of just how, exactly, the internal nodes or connections were supposed to function as representational states, I failed to receive a satisfactory answer. Often my queries would be met with a shrug and reply of "what else could they be doing?" It seemed the default assumption was that these hypothetical internal structures must be representations and that the burden of proof was upon anyone who wished to deny it. I first expressed my concerns about the explanatory value of connectionist representations much later, in a paper published in *Mind and Language* (Ramsey, 1997). At the time, William Bechtel correctly noted that my arguments, if they worked, would challenge not only the notions of representation associated with connectionism, but also the representational posits associated with a much wider range of theories. Although Bechtel intended this point as a problem with my view, I saw it as revealing a serious problem with the way people were thinking about representation within the broader cognitive science community.

Since that time, my skepticism about popular conceptions of representation has only grown, though not entirely across the board. I have also come to appreciate how some notions of representation actually do succeed in addressing my worries about representational function. To be sure, these notions of representation have their problems as well. But as the saying goes, there are problems and then there are *problems*. My belief is that some of the notions of representation we find in cognitive research need a little fixing up here and there, whereas other notions currently in vogue are hopeless non-starters. As it happens, the notions of representation that I think are promising are generally associated with the classical computational theory of cognition, whereas the notions I think are non-starters have been associated with the newer, connectionist and neurologically-based theories. Spelling all this out is one of the main goals of this book. The central question my analysis will ask is this: "Do the states characterized as representation in explanatory framework X actually serve as representations, given the processes and mechanisms put forth?" The answer I'm going to offer is, by and large, "yes" for the classical approach, and "no" for the newer accounts. When we look carefully at the way the classical framework explains cognitive processes, we find that talk of representation is justified, though this justification has been clouded in the past by misguided analyses. However, when we look at the explanatory strategies provided by the newer accounts, we find something very different. Although neuroscientific and connectionist theories characterize states and structures as inner representations, there is, on closer inspection, no compelling basis for this characterization.

It might be assumed that such an assessment would lead to an endorsement of the classical framework over the newer accounts. But that would follow only if we presume that psychological theories absolutely must invoke representational states in their explanations of cognitive capacities. I think it is an open empirical question whether or not the brain actually uses representational states in various psychological processes. Most of the theories I criticize here still might prove workable, once the conceptual confusions about representation are cleared away. What my analysis does reveal, however, is that something very interesting is taking place in cognitive science. When new scientific theories are offered as alternatives to more established views, proponents of the new perspective are sometimes reluctant to abandon the familiar notions of the older framework, even when those posits have no real explanatory role in the new accounts. When this happens, the old notions may be re-worked as theorists contrive to fit them into an explanatory framework for which they are ill-suited. One of the central themes of this book is that something very much like this is currently taking place in cognitive science. My claim is that the representational perspective, while appropriate for classical computational cognitive science, has been carried over and assigned to new explanatory frameworks to which it doesn't actually apply. Although investigators who reject the classical framework continue to talk about internal representations, the models and theories many of them propose neither employ, nor need to employ, structures that are actually playing a representational role. I will argue that cognitive research is increasingly moving away from the representational paradigm, although this is hidden by misconceptions about what it means for something to serve as a representational state.

Thus, my primary objective is to establish both a positive and a negative thesis. The positive position is that, contrary to claims made by critics of conventional computationalism, the classical framework does indeed posit robust and explanatorily valuable notions of inner representation. To see this, we need to abandon what I call the "Standard Interpretation" of computational symbols as belief-like states, and instead view them as representations in a more technical sense. Computational explanation often appeals to mental models or simulations to account for how we perform various cognitive tasks. Computational symbols serve as elements of such models, and, as such, must *stand in for* (i.e., represent) elements or aspects of that which is being modeled. This is one way in which the classical picture employs a notion of representation that is doing real explanatory work. My negative claim is that the notions of representation invoked by many non-classical accounts of cognition do not have this sort of explanatory value. Structures that are described as representations are actually playing a functional role that, on closer inspection, turns out to have little to with do anything recognizably representational in nature. For example, proposed structures are often characterized as representations because they faithfully respond to specific stimuli, and in turn causally influence other states and processes. My claim will be that this is not a representational role, and that these posits are better described as relay circuits or causal mediators.

In arguing for both the positive and negative theses, I will appeal to what I call the "job-description challenge." This is the challenge of explaining how a physical state actually fulfills the role of representing in physical or computational process – accounting for the way something actually *serves* as a representation in a cognitive system. In the philosophy of psychology, the emphasis upon content has led many to assume that a theory of content provides a theory of representation. But an account of content is only one part of the story. The question of how a physical structure comes to function as a representation is clearly different from (though related to) the question of how something that is presumed to function as a representation comes to have the intentional content it does. I claim that when we take the former question seriously, we can see that, by and large, classical computational representations meet the job-description challenge, but the notions of representation in the newer theories do not.

The analysis I will offer here is inspired by Robert Cummins's suggestion that the philosophy of psychology (and the philosophy of representation in particular) should primarily be an enterprise in the philosophy of science. Just as philosophers of physics might look at the explanatory role of the posits of quantum physics, or a philosopher of biology might look at different conceptions of genes, my agenda is to critically examine the different ways cognitive scientists appeal to notions of representation in their explanations of cognition. I believe such an assessment reveals that cognitive science has taken a dramatic anti-representational turn that has gone unnoticed because of various mis-characterizations of the posits of the newer theories. Cognitive theories are generally described as distinct from behaviorist accounts because they invoke inner representation. However, if many current cognitive theories are, as I argue, not actually representational theories, then we need to reconsider the scope of the so-called "cognitive revolution" and the degree to which modern cognitivism is really so different from certain forms of behaviorism. Moreover, a non-representational psychology would have important implications for our commonsense conception of the mind – our so-called "folk psychology." Since commonsense psychology is deeply committed to mental representations in the form of beliefs and other propositional attitudes, this non-representational reorientation of cognitive science points in the direction of eliminative materialism – the radical thesis that beliefs don't actually exist. Eliminativism would bring about a cataclysmic shift in our understanding not just of psychological processes, but in our overall conception of ourselves. Thus, the developments that I will try to illuminate here are of enormous significance, despite having gone unnoticed by most cognitive scientists and philosophers of psychology.

To show all this, the book will have the following structure. In the first chapter, I introduce some of the issues and concerns that will take center stage in the subsequent chapters. After explaining the central goals of the book, I look at two families of representational concepts – one mental, the other non-mental – to get a preliminary handle on what it might mean to invoke representations as explanatory posits in cognitive science. I argue that our commonsense understanding of representation constrains what can be treated as a representation and presents various challenges for any scientific account of the mind that claims to be representational in nature. I also introduce the job description challenge and argue that theories that invoke representations carry the burden of demonstrating just how the proposed structure is supposed to serve as a representation in a physical system. Moreover, I argue this must be done in such a way that avoids making the notion of representation completely uninteresting and divorced from our ordinary understanding of what a representation actually is.

The goal of the second chapter is to present what I take to be a popular set of assumptions and tacit attitudes about the explanatory role of representation in the classical computational theory of the mind. I'll suggest that these assumptions and attitudes collectively give rise to an outlook on representation that amounts to a sort of merger between classical computational theory and folk psychology. This has led to a way of thinking about computational representations that suggests their primary explanatory function is to provide a scientific home for folk notions of mental representations like belief. I call this the "Standard Interpretation" of classical computationalism. After spelling out what I think the Standard Interpretation involves, I'll try to show that it leads us down a path where, despite various claims to the contrary, we wind up wondering whether the symbols of classical models should be viewed as representations at all. This path has been illuminated by two important skeptics of classical AI, John Searle and Stephen Stich. Searle and Stich both exploit the alleged link between classicalism and folk psychology to challenge the claim that the classical framework can or should appeal to inner representations. I'll present Searle's and Stich's criticism of representationalism and examine the ways defenders of the Standard Interpretation have responded. In the final analysis, I'll argue the Standard Interpretation leaves in doubt the representational nature of computational states.

In the third chapter, I reject the Standard Interpretation and provide what I believe is the proper analysis of representation in the classical computational theory. Picking up on themes suggested by prior writers (such as John Haugeland and Robert Cummins), I argue that there are two related notions playing valuable explanatory roles, and that neither notion is based upon commonsense psychology. One notion pertains to the classical computational strategy of invoking inner computational operations to explain broader cognitive capacities. I argue that these inner subcomputations require inputs and outputs that must be representational in nature. The second notion, designated as "S-representation," pertains to data structures that in classical explanations serve as elements of a model or simulation. That is, according to many theories associated with the classical framework, the brain solves various cognitive problems by constructing a model of some target domain and, in so doing, employs symbols that serve to represent aspects of that domain. After providing a sketch of each notion, I consider two popular criticisms against them and argue that both criticisms can be handled by paying close attention to the way these notions are actually invoked in accounts of cognition. Finally, I address a number of side issues associated with these notions, such as their explanatory connection to computational rules and the question of whether they would vindicate the posits of folk psychology.

The fourth chapter begins the negative phase of the book and is devoted to exploring what I call the "receptor" notion of representation that appears in a wide range of theories in cognitive neuroscience and connectionist modeling. This style of representation often borrows from Shannon and Weaver's theory of information, and rests on the idea that neural or connectionist states represent certain stimuli because of a co-variance or nomic dependency relation with those stimuli. The work of Fred Dretske provides what is perhaps the clearest and most sophisticated defense of the explanatory value of this family of representational notions. However, despite Dretske's impressive support for this type of representation, I argue that the notion is too weak to have any real explanatory value. What gets characterized as a representation in this mold is often playing a functional role more akin to a non-representational relay circuit or simple causal mediator. In these cases, any talk of "information carrying" or representational content could be dropped altogether without any real explanatory loss. I look closely at the arguments presented by Dretske and suggest that his account of representation is inadequate because it fails to meet the job description challenge.

The fifth chapter looks at a somewhat scattered family of representational notions found in various accounts of neurological processes, artificial intelligence and in various connectionist networks. Here the basic idea is that the functional architecture of a system plays a representational role largely because it is causally relevant to the production of various types of output. I characterize this as the "tacit" notion of representation since there is typically no one-to-one mapping between cognitive structures and individually represented items. The functional architecture of a system is said to encode information holistically, and this is thought to serves as the system's "know-how." After explaining the core features associated with this family of representational notions, I offer a critical evaluation and argue that, like the receptor notion, it fails to meet the job description challenge. Once again, representation is confused with something else; in this case, with the dispositional properties of the underlying architecture. Since there is no real motivation for treating these sorts of structures as representations, I defend the position that we should stop thinking of them in this way.

The sixth and final chapter addresses three important topics related to my analysis. First, to solidify my earlier claims, I offer a more direct comparison between the receptor and S-representational notions in the form of imaginary, quasi-robotic systems attempting to navigate a track. My aim here is to make clearer just how and why the receptor notion runs into trouble, while the S-representation notion is better suited for psychological theorizing. Second, in recent years, pockets of anti-representationalism have developed in various areas such as robotics research and Dynamic Systems Theory, and defenders of representationalism have offered a number of intriguing responses to these challenges. Because some of these defenses of representation can also be seen as challenging some of my own skeptical claims, it is important to examine them closely to see if they rescue the representational posits from my critique. I argue that they fail to do this, and that if anything they help show just why certain notions are ill-suited for cognitive modeling. Finally, I address some of the ramifications of the arguments presented in the earlier chapters. If many representational notions now employed in cognitive research are, as I suggest, not representational at all, then we need to rethink the extent to which these newer accounts are really so different from the "pre-cognitivist," behaviorist theories of psychological processes. I suggest that some behaviorists, like Hull, often proposed internal mediational states that were not significantly different, in terms of functionality, from what today gets described in representational terms. A second implication of my arguments concerns the status of folk psychology. If I'm right, then many models of cognitive processes currently being proposed do not actually appeal to inner representational states. Because commonsense psychology is deeply committed to mental representations, the truth of these theories would entail eliminative materialism, the radical thesis that folk psychology is fundamentally wrong and states like beliefs and desire do not actually exist. In the final section of this chapter, I'll sketch one way this might come about that is not as preposterous as it initially sounds.

This book has taken a long time to complete and I have received a great deal of help along the way from numerous colleagues, students and friends. Among those providing helpful criticisms, insights and suggestions are William Bechtel, Tony Chemero, Marian David, Neil Delaney, Michael Devitt, Steve Downes, Chris Eliasmith, Keith Frankish, Carl Gillett, Terry Horgan, Todd Jones, Lynn Joy, Matthew Kennedy, Jaegwon Kim, John Schwenkler, Matthias Scheutz, Peter Godfrey-Smith, Stephen Stich, and Michael Strevens. I'm especially grateful to Robert Cummins, Fred Dretske, Keith Frankish, Tony Lambert, Leopold Stubenberg, Fritz Warfield, and Daniel Weiskopf who read substantial portions of earlier drafts of the manuscript and provided extremely helpful suggestions. I also want to thank Ryan Greenberg and Kate Nienaber who did the illustrations that appear in the final chapter, and my sister, Julie Talbot, who rendered some much-needed proofreading of the entire manuscript. Hilary Gaskin of Cambridge University Press provided everything an author can hope for from an editor, and Susan Beer made the copy-editing remarkably simple and straightforward. I should also acknowledge the many climbing partners who over the years, on endless drives and at cramped belay stances, humored me as I tried out some of the ideas that appear here – I imagine that occasionally one or two considered cutting the rope.

Some of the arguments presented here have appeared in a different context in other published works, most notably in "Are Receptors Representations?" (2003, *Journal of Experimental and Theoretical Artificial Intelligence* 15: 125–141); "Do Connectionist Representations Earn Their Explanatory Keep?" (1997, *Mind and Language* 12 (1): 34–66), and "Rethinking Distributed Representation" (1995, *Acta Analytica* 14: 9–25). I have also benefited a great deal from feedback from audiences at the University of Utah, the University of Cincinnati, The University of Nevada, Las Vegas, the University of Notre Dame, the Southern Society of Philosophy and Psychology Annual Meeting (2005, Durham, NC); Cognitive Science in the New Millennium Conference (2002, Cal. State Long Beach), Society for Psychology and Philosophy Annual Meeting (1994, Memphis, Tennessee), and the IUC Conference on Connectionism and the Philosophy of Mind (1993, Bled, Slovenia). I am extremely grateful to the University of Notre Dame for awarding me with an Associative Professor's Special Leave to complete this book. I would also like to thank my department chair, Paul Weithman, who has been especially supportive of this project in a variety of different ways.

Finally, I would like to offer a special thanks to Stephen Stich, whose support and advice over the years has always proven invaluable. Nearly twenty-five years ago, he presented a devastating challenge to the received view that cognitive processes require mental representations (Stich 1983). Since no other person has had as much of an impact on my philosophical career, it is perhaps not surprising that, despite significant changes in cognitive research and the philosophy of mind, I find myself a quarter century later promoting views that are in much the same skeptical spirit.

1

Demands on a representational theory

A common feature of scientific revolutions is the discarding of the theoretical posits of the older theory in favor of the posits invoked by the new theory. The abrupt shift in the theoretical ontology is, of course, one of the things that can make a scientific upheaval so dramatic. Sometimes, however, it happens that the displaced posits hang around for a considerable stretch of time. Despite losing their explanatory value, they nevertheless retain their stature and prominence as even revolutionary thinkers resist abandoning something central to their basic understanding of the subject. The posit is perhaps transformed and re-worked as theorists contrive to fit it into a new explanatory framework for which it is ill-suited. Yet its appearance in the new theory is motivated not by any sort of explanatory necessity, but by a reluctance to reject familiar ontological commitments. When this happens, there can be a number of undesirable consequences. One is a failure to appreciate just how radical the new theoretical framework is; another is a confused understanding of the explanatory framework of the new theory, due to an extended attempt to incorporate theoretical posits that don't belong.

The status of celestial spheres shortly after the Copernican revolution helps illustrate this point. In Ptolemy's system, the spheres did real explanatory work; for instance, they helped explain what kept the massive array of stars in place as they orbited around the Earth. Without some sort of "starry vault" to anchor the stars as they rotated, they would inevitably lose their relative positions and we would look up to a different sky every night. The solid spheres provided the secure medium to prevent this from happening. But with the new Copernican cosmology, the stars stopped moving. Instead, it was the Earth that rotated, spinning on a 24-hour cycle and creating the false impression of revolving stars. Consequently, a central assumption that supported the need for celestial spheres was dropped from the new model, and it became possible to view the stars as stationary points in empty space. And yet, Copernicus and others refused to abandon the idea of semi-solid spheres housing not only the stars, but the different planets as well. This reluctance to discard the spheres from the new cosmology was no doubt due to considerations that went substantially beyond science. Historical, theological, cultural, and perhaps even "folk" considerations all played an important role in preserving the spheres, despite increasing problems in making them conform to the new theory. Although Tycho Brahe recommended abandoning solid spheres, Kepler rescued them as semi-abstract posits that he felt were essential for understanding the celestial system. It wasn't until Descartes's re-conceived space as a giant container that people let go of the idea of a starry vault (Crowe 2001; Donahue 1981).

The central theme of this book is that something very similar is currently taking place in our scientific understanding of the mind. In cognitive science, there has been something like a central paradigm that has dominated work in psychology, linguistics, cognitive ethology and philosophy of mind. That paradigm is commonly known as the classical computational theory of cognition, or the CCTC for short.[1] At the heart of the classical paradigm is its central explanatory posit – internal symbolic representations. In fact, the notion of internal representation is the most basic and prevalent explanatory posit in the multiple disciplines of cognitive science. The representational underpinning of cognitive science is, as one author puts, "what the theory of evolution is to all of biology, the cell doctrine to cellular biology, the notion of germs to the scientific concept of disease, the notion of tectonic plates to structural geology" (Newell 1980, p. 136). In the minds of many psychologists, linguists, ethologists and philosophers, the positing of internal representations is what *makes* a given theory cognitive in nature.

However, in the last two decades there have been several radical theoretical departures from the classical computational account. Connectionist modeling, cognitive neuroscience, embodied cognitive accounts, and a host of other theories have been presented that offer a very different picture of the architecture and mechanisms of the mind. With new processes like "spreading activation," "distributed constraint satisfaction," and "stochastic-dynamical processes," the operations of what John Haugeland (1997) has referred to as "new fangled" AI systems don't have much in common

[1] It is also sometimes called "GOFAI" for "Good-Old-Fashioned-Artificial-Intelligence," the "Physical Symbol Hypothesis," the "Computer Model of the Mind" (CMM), "Orthodox Computationalism," the "Digital Computational Theory of Mind" (DCTM), and a host of other names. There are now so many labels and acronyms designating this class of theories that it is impossible to choose one as "the" accepted name.

with the familiar symbol-based approach of the classical paradigm. Yet despite massive differences between classical accounts and the newer theories, the latter continue to invoke inner representations as an indispensable theoretical entity. To be sure, the elements of the newer theories that are characterized as representations look and act very differently than the symbols in the CCTC. Nevertheless, the new accounts share with conventional computational theories the basic idea that inner structures in some way serve to stand for, designate, or mean something else. The commitment to inner representations has remained, despite the rejection of the symbol-based habitat in which the notion of representation originally flourished.

My aim is to argue that this is, for the most part, a mistake. A central question I'm going to address in the following pages is, "Does the notion of inner representation do important explanatory work in a given account of cognition?" The answer I'm going to offer is, by and large, "yes" for the classical approach, and "no" for the newer accounts. I'm going to suggest that like the notion of a starry vault, the notion of representation has been transplanted from a paradigm where it had real explanatory value, into theories of the mind where it doesn't really belong. Consequently, we have accounts that are characterized as "representational," but where the structures and states called representations are actually doing something else. This has led to some important misconceptions about the status of representationalism, the nature of cognitive science and the direction in which it is headed. It is the goal of this book to correct some of these misconceptions.

To help illustrate the need for a critical analysis like the one I am offering, try to imagine what a non-representational account of some cognitive capacity or process might look like. Such a thing should be possible, even if you regard a non-representational account as implausible. Presumably, at the very least, it would need to propose some sort of internal processing architecture that gives rise to the capacity in question. The account would perhaps invoke purely mechanical operations that, like most mechanical processes, require internal states or devices that in their proper functioning go into particular states when the system is presented with specific sorts of input. But now notice that in the current climate, such an account would turn out to be a representational theory after all. If it proposes particular internal states that are responses to particular inputs, then, given one popular conception of representation, these would qualify as representing those inputs. And, according to many, any functional architecture that is causally responsible for the system's performance can be characterized as encoding the system's knowledge-base, as implicitly representing the system's know-how. If we accept current attitudes about the nature of cognitive representation, a non-representational, purely mechanistic account of our mental capacities is not simply implausible – it is virtually *inconceivable.* I take this to be a clear indicator that something has gone terribly wrong. The so called "representational theory of mind" should be an interesting empirical claim that may or may not prove correct; representations should be unique structures that play a very special sort of role. In many places today, the term "representation" is increasingly used to mean little more than "inner" or "causally relevant" state.

Returning for a moment to our analogy between celestial spheres and representation, it should be noted that the analogy is imperfect in a couple of important ways. First, in the case of the spheres, astronomers had a fairly good grasp of why they were needed in Ptolemy's system. By contrast, there has been much less clarity or agreement about the sort of role the notion of representation plays in cognitive science theories in general, including the older paradigm. Thus, one of my chores will be to sort out just how and why such a notion is needed in the CCTC. A second dis-analogy is that in the case of the spheres, there was, for the most part, a single notion at work and it was arguably that same notion that found its way into Copernicus's system. However, in the case of representation, there are actually clusters of very distinct notions that appear in very distinct theories. Most of these notions are based on ideas that have been around for a long time and certainly pre-date cognitive science. Some of these notions, when embedded in the right sort of account of mental processes, can play a vital role in the theory. Other notions are far more dubious, at least as explanatory posits of how the mind works. My claim will be that, for the most part, the notions that are legitimate – that is, that do valuable explanatory work – are the ones that are found in the CCTC. The notions of representation that are more questionable have, by and large, taken root in the newer theories. I propose to uproot them.

Methodological matters

The goals of this book are in many ways different from those of many philosophers investigating mental representation. For some time philosophers have attempted to develop a naturalistic account of intentional content for our commonsense notions of mental representation – especially our notion of belief. By "naturalistic account" I mean an account that explains the meaningfulness of beliefs in the terms of the natural sciences, like physics or biology. The goal has been to show how the representational character of our beliefs can be explicated as part of the natural world. While many of these accounts are certainly inspired by the different ways researchers appeal to representation in cognitive theories, they neither depend upon nor aim to enhance this research. Instead, the work has been predominantly conceptual in nature, and the relevant problems have been of primary interest solely to philosophers.

By contrast, my enterprise should be seen as one based in the philosophy of science – in particular, the philosophy of cognitive science. The goal will be to explore and evaluate some of the notions of representation that are used in a range of cognitive scientific theories and disciplines. Hence, the project is similar to that, say, of a philosopher of physics who is investigating the theoretical role of atoms, or a philosopher of biology exploring and explicating competing conceptions of genes. This way of investigating mental representation has been explicitly adopted and endorsed by Robert Cummins (1989) and Stephen Stich (1992). Cummins's explanation of this approach is worth quoting at length:

> It is commonplace for philosophers to address the question of mental representation in abstraction from any particular scientific theory or theoretical framework. I regard this as a mistake. Mental representation is a theoretical assumption, not a commonplace of ordinary discourse. To suppose that "commonsense psychology" ("folk psychology"), orthodox computationalism, connectionism, neuroscience, and so on all make use of the same notion of representation is naive. Moreover, to understand the notion of mental representation that grounds some particular theoretical framework, one must understand the explanatory role that framework assigns to mental representation. It is precisely because mental representation has different explanatory roles in "folk psychology," orthodox computationalism, connectionism, and neuroscience that it is naive to suppose that each makes use of the same notion of mental representation. We must not, then, ask simply (and naively) "What is the nature of mental representation?"; this is a hopelessly unconstrained question. Instead, we must pick a theoretical framework and ask what explanatory role mental representation plays in that framework and what the representation relation must be if that explanatory role is to be well grounded. Our question should be "What must we suppose about the nature of mental representation if orthodox computational theories (or connectionist theories, or whatever) of cognition are to turn out to be true and explanatory?" (1989, p. 13)

Cummins's own analysis of representation in classical computational theory will be discussed in some detail in chapter 3, where I will offer modifications to his account. For now, I want to appeal to the Cummins model to make clear how my own account should be understood. My analysis is very much in the same spirit as what Cummins suggests, but with a couple of caveats. First, Cummins and Stich seem to assume that to demarcate the different notions of representation one should focus upon the theory in which the notion is embedded. However, a careful survey of cognitive research reveals that the same core representational notions appear in different theories and different disciplines. Hence, a better taxonomy would be one that cuts across different theories or levels of analysis and classifies types of representational notions in terms of their distinctive characteristics. Toward the end of this chapter, I'll explain in more detail the demarcation strategy I plan to use. Second, Cummins doesn't mention the possibility that our deeper analysis might discover that the notion of representation invoked in a theory actually turns out to play *no* explanatory role. Yet I'll be arguing that this is precisely what we do find when we investigate some of the more popular accounts of cognition commonly characterized as representational in nature.

Because the expanse of cognitive science is so broad, my analysis cannot be all-encompassing and will need to be restricted in various ways. For instance, my primary focus will be with theories that attempt to explain cognition as something else, like computational or neurological processes. In such theories, researchers propose some sort of process or architecture – a classical computational system or a connectionist network – and then attempt to explain cognition by appealing to this type of system. In these accounts, talk of representation arises when structures inherent to the specific explanatory framework, like data structures or inner nodes, are characterized as playing a representational role. Theories of this sort are reductive in nature because they not only appeal to representations, but they identify representations with these other states or structures found in the proposed framework. This is to be contrasted with psychological theories that appeal to ordinary notions of mental representation without pretending to elaborate on what such representation might be. For example, various theories simply presuppose the existence of beliefs and concepts to account for different dimensions of the mind, offering no real attempt to further explain the nature of such states, or representation in general. I'll be more concerned with theories that invoke representations as part of an explanatory system and at the same time offer some sense of what internal representations actually are.

Since my aim is to assess critically the notion of representation in cognitive theories, I won't be arguing for or against these theories themselves, apart from my evaluation of how they use a notion of representation. The truth or falsehood of any of these theories is, of course, an empirical matter that will depend mostly on future research. Even when I claim that a cognitive theory employs a notion of representation that is somehow bogus, or is treating structures as representations that really aren't, I don't intend this to suggest that the theory itself is utterly false. Instead, I intend it to suggest that the theory needs conceptual re-working because it is mis-describing a critical element of the system it is trying to explain.

Still, even this sort of criticism raises an important question about the role of philosophy in empirical theory construction. Why should a serious cognitive scientist who develops an empirical theory of cognition that employs a notion of representation pay attention to an outsider claiming that there is something wrong with the notion of representation invoked? What business does a philosopher have in telling any researcher how to understand his or her own theory? My answer is that in the cross-disciplinary enterprise of cognitive science, what philosophers bring to the table is a historical understanding of the key notions like representation, along with the analytic tools to point out the relevant distinctions, clarifications, implications, and contradictions that are necessary to evaluate the way this notion is used (and ought not to be used). To some degree, our current understanding of representation in cognitive science is in a state of disarray, without any consensus on the different ways the notion is employed, on what distinguishes a representational theory from a non-representational one, or even on what something is supposed to be doing when it functions as a representation. As psychologist Stephen Palmer notes, "we, as cognitive psychologists, do not really understand our concepts of representation. We propose them, and talk about them, argue about them, and try to obtain evidence in support of them, but we do not understand them in any fundamental sense" (Palmer 1978, p. 259). It is this understanding of representation, in a fundamental sense, that philosophers should help provide.

One reason for the current state of disorder regarding representation is that it is a theoretical posit employed in an unusually broad range of disciplines, including the cognitive neurosciences, cognitive psychology, classical artificial intelligence, connectionist modeling, cognitive ethology, and the philosophy of mind and psychology. This diversity multiplies when we consider the number of different theories within each of these disciplines that rely on notions of representation in different ways. It would be impossible to examine all of these different theoretical frameworks and applications of representational concepts. Hence, the overall picture I want to present will need to be painted, in spots, with broad strokes and I'll need to make fairly wide generalizations about theories and representational notions that no doubt admit of exceptions here and there. This is simply an unavoidable part of doing this type of philosophy of science, given the goal of providing general conclusions about a diverse array of trends and theories on this topic. If what I say does not accurately describe your own favorite theory or model, I ask that you consider my claims in light of what you know about more general conventions, attitudes, assumptions and traditions.

If I am going to establish that certain notions of representation in cognitive science are explanatorily legitimate while others are not, we need to try to get a better sense of what constitutes "explanatory legitimacy." Given the current lack of agreement about representation, figuring out just how such a notion is supposed to work in a theory of mental processes is far from easy. Despite the large amount of material written on mental representation over the years, it is still unclear how we are supposed to think about it. As John Searle once noted, "There is probably no more abused a term in the history of philosophy than 'representation' ..." (1983, p. 11). Arguably, the same could be said about "representation" in the history of cognitive science. What does the positing of internal representations amount to? When is it useful to do so and when is it not? Exactly what is being claimed about the mind/brain when it is claimed to have representational states? Answering these questions is, in large measure, what this book will try to do. As a first pass, it will help to first step back and consider in more general terms some of our ordinary assumptions and attitudes about representational states.

1.1 REPRESENTATION AS CLUSTER CONCEPT(S)

Cognitive researchers often characterize states and structures as representations without a detailed explication of what this means. I suspect the reason they do this is because they assume they are tapping into a more general, pre-theoretical understanding of representation that needs no further explanation. But it is actually far from clear what that ordinary conception of representation involves, beyond the obvious, "something that represents." Perhaps the first thing we need to recognize is that, as others have pointed out (Cummins 1989; von Eckardt 1993), it is a mistake to search for *the* notion of representation. Wittgenstein famously suggested that concepts have a "family-resemblance" structure, and to demonstrate his point, he invoked the notoriously disjunctive notion of a game. But Wittgenstein could have just as easily appealed to our ordinary notion of representation to illustrate what he had in mind. We use the term "representation" to characterize radically different things with radically different properties in radically different contexts. It seems plausible that our notion of representation is what is sometimes called a "cluster" concept (Rosch and Mervis 1975; Smith and Medin 1981) with a constellation of different types that share various nominal features, but with no real defining essence. If this is the case, then one popular philosophical strategy for exploring representation in cognitive science is simply untenable.

When trying to understand representation in cognitive science, writers often offer semi-formal, all-encompassing definitions that are then used as criteria for determining whether or not a theory invoking representations is justified in doing so. Initially, this might seem like a perfectly reasonable way to proceed. We can simply compare the nature of the posit against our crisp definition and, with a little luck, immediately see whether the alleged representation makes the cut. However, I believe this strategy has a number of severe flaws. First, in many cases the definition adds more mystery and confusion that it clears away. For example, Newell has famously defined representation in terms of a state's capacity to designate something else, and then defines designation in this way: "An entity X designates an entity Y relative to a process P, if, when P takes X as input, its behavior depends on Y" (1980, p. 156).

It is far from clear how this definition is supposed to refine our understanding of designation or representation. After all, my digestive processes sometimes takes a cold beer as input and when it does so its behavior often depends on whether or not I've had anything else to eat, along with a variety of other factors. Does this mean a cold beer designates my prior food intake? Presumably not, yet it appears the definition would say that it does. Newell clearly intends to capture a relation between X, P and Y that is different from this, yet the definition fails to explicate what this relation might be.

Second, virtually all of the definitions that have been offered give rise to a number of intuitive counter-examples. As we have just seen, Newell's criteria, taken as sufficient conditions, would suggest that a beer I've ingested serves a representational function, which it clearly does not. As we will see in the forthcoming chapters, similar problems plague the definitions offered by other writers who propose definitions of representation. Counter-examples come in two forms – cases that show a proposed definition is too inclusive (i.e., treat non-X as if they are Xs) and cases that show a proposed definition is too exclusive (i.e., treat actual Xs as if they are not Xs). Definitions of representation typically fail because of the former sort of counter-examples – states and structures that play no representational role are treated as if they actually do.

Now it might be thought that these difficulties are simply due to a bunch of flawed definitions, while the original goal of constructing a general definition for representation is still worth pursing. Yet the research on categorization judgments suggests there is reason to think these problems run deeper and are symptomatic not of bad analysis, but of the nature of our underlying pre-theoretical understanding of representation. If Rosch and various other psychologists are correct about the disjunctive way we encode abstract concepts, then the difficulties we see with these definitions are exactly what we should expect to find. Simple, tidy, conjunctive definitions will always fall short of providing a fully satisfactory or intuitive analysis. They might capture one or two aspects of some dimension of our general understanding, but they won't reveal the multi-faceted nature of how we really think about representation.

Suppose these psychologists are right about our conceptual machinery and that our concept of representation is itself a representation of an array of features clustered around some sort of prototype or group of prototypes. This would make any crisp and tidy definition artificial, intuitively unsatisfying, and no better than a variety of other definitions that would generate very different results about representation in theories of the mind. If we want to evaluate the different notions of representation posited in scientific theories, a more promising tack would be to carefully examine the different notions of representation that appear in cognitive theories, get as clear as possible about just what serving as a representation in this way amounts to, and then simply ask ourselves – is this thing really functioning in a way that is recognizably representational in nature? In other words, instead of trying to compare representational posits against some sort of contrived definition, we can instead compare it directly to whatever complex concept(s) we possess to see what sort of categorization judgment is produced. If, upon seeing how the posit in question actually functions we are naturally inclined to characterize its role as representational in nature, then the posit would provide us with one way of understanding how physical systems can have representations. If, on the other hand, something is functioning in a manner that isn't much like what we would consider to be a representational role, then the representational status of the posit, along with its embedding theory, is in trouble. This is roughly how my analysis will proceed – by exploring how a representational posit is thought to operate in a system, and then assessing this role in terms of our ordinary, intuitive understanding of what a representation is and does. To some degree, this means our analysis will depend on a judgment call. If this is less tidy than we would like, so be it. I would prefer a messier analysis that presents a richer and more accurate account of representation than one that is cleaner but also off the mark. Eventually, we may be able to construct something like a general analysis or theory of representation. But this can only happen after first exploring the ways in which physical structures may or may not accord with our more basic, intuitive understanding of representation.

A very different question worth considering is this: why should we care if a given representational posit accords with our commonsense understanding of representation in the first place? If these are technical, scientific posits, what difference does it make whether the theorist uses the term "representation" to refer to things that behave in a manner sanctioned by intuition? Isn't is really just the explanatory value of a theoretical posit that matters? And if so, isn't it trivially true that cognitive systems use representations? An illustration of this attitude is provided by Roitblat (1982), who happily proclaims that, "[t]o assume the existence of a representation is rather innocuous and should rarely be an issue for theoretical dispute" (Roitblat 1982, p. 355). Since Roitblat defines representation as *any* internal change caused by experience, it is not surprising that he thinks it pointless to wonder about their existence.

However, I actually think that quite a lot rides on whether or not a representational posit actually functions in a way that we are naturally inclined to recognize as representational. First, it is important to think carefully about what it means to say that a given notion is doing important explanatory work. Suppose someone claims to have a representational theory of diseases, and posits representational states as the cause of most illnesses. Upon further analysis, we discover that the theorist is simply using the term "representation" to refer to ordinary infectious agents, like viruses and bacteria. Moreover, we discover that there is nothing intuitively representational about the role the theory assigns to these agents – they just do the things infectious agents are ordinarily assumed to do. Notice how silly it would be for the theorist to defend his representational account by pointing out that he isn't interested in our ordinary notion of representation, and that what matters is that his representational posits do important explanatory work. While the posits would indeed do explanatory work, they wouldn't actually be serving as *representational* posits. This would not be a case where a technical notion of representation is playing some explanatory role. Instead, this would be a scenario where a notion of representation would not be playing *any* explanatory role; it would be completely absent from the theory. All of the work would be done by ordinary notions of infectious agents. This is because there is nothing about the job these states are doing that is intuitively recognizable as representational in nature. Unless a posit is in *some* way grounded in our ordinary understanding of representation, it is simply not a representational posit, in any sense.

In earlier work (Ramsey 1997), I chose not to address the issue of whether or not a proposed form of representation actually was a representation, and instead focused on the question of explanatory utility, asking if a notion of representation did any explanatory work. I now think this was a mistake. It was a mistake because what matters is not explanatory work, but explanatory work *qua* representation. In showing that a posit fails to do explanatory work *qua* representation, what is typically shown is that the proposed posit doesn't function in a representational manner; that is, it is not a representation after all. So the metaphysical issues cannot be avoided, even if one's primary interest is with questions of explanatory utility. There are different ways a theoretical posit from an older framework can be mistakenly retained in a new framework. One way, suggested by the case of the crystal spheres, is if the posit fails to correspond to anything in the new ontology. But another way is if some part of the new ontology is characterized as playing the role associated with the old posit when in truth, it is playing a completely different role. That is what I will claim is happening with the notion of representation.

Second, the positing of inner representations typically comes with a lot of assumptions, expectations, concerns, inferential entitlements, and other theoretical attachments that are rooted in (and licensed by) our ordinary ways of thinking about representation. The significance of a representational theory of the mind stems in large measure from the different elements that are associated with representational states as ordinarily understood. For example, when theorists posit inner representations, they typically assume that they now have an important way to explain how the system can fail to behave appropriately. It is now possible to explain faulty behavior as sometimes stemming from false representations of the world. In fact, considerable philosophical effort has been devoted to explaining how it is actually possible for a physical state to be in error – to misrepresent the nature of reality. This is an important topic because the possibility of misrepresentation is built into our ordinary way of understanding what it is to represent. If someone announced that they were using a technical notion of representation that didn't admit of misrepresentation, we would not think that this is just another way of handling the problem of error. Instead, we would think that whatever the posited state was doing, it wasn't playing a representational role. We can't posit representational states to do many of the things they are supposed to do in a theory unless the posit itself is sufficiently similar to the sort of things we pre-theoretically think representations are.

This last point also helps us see that there is more at stake here than a mere terminological or semantic squabble. With a simple terminological mistake, a non-A is mistakenly called an "A," though it is not ascribed any of the features normally associated A. This might happen when someone is learning a language. In the case of real conceptual confusion, on the other hand, a non-A is called an "A" and also treated as having all (or most) of the features normally associated with A. It is clearly one thing to mistakenly think the word "dog" refers to cats, it is quite another thing to mistakenly think that dogs are a type of cat. The confusion I will be addressing involves the latter sort of mistake – people thinking that non-representational states and structures really are a type of representation. This leads them to make the further mistake of thinking that the sort of conceptual linkages and accompaniments associated with representation should be ascribed to non-representational entities.

Finally, contrary to Roitblat's claim, the question of whether or not the brain performs cognitive tasks by using inner representations is an important one that deserves to be investigated with the same seriousness that we investigate other important empirical questions. Notice how many traditional problems could be resolved by just ignoring our intuitive understanding of things and instead offering new definitions. Can machines be conscious? Well, let's just define consciousness as electrical activity and thereby prove that they can. Do non-human primates communicate with a language? Sure, if we think of language as any form of communication. Does smoking really cause lung cancer? No, not if we ignore our ordinary way of thinking about causation and employ a technical notion where to be a cause is to be a necessary and sufficient condition. Most of us would treat this sort of strategy for addressing these questions as uninteresting ploys that dodge the real issues. Similarly, any suggestion that we should answer the question, "does this system employ inner representations?" in a manner that ignores our intuitive understanding of what a representation is and does is equally misguided. Of course, this doesn't mean that there can't be notions of representation that are somewhat technical, or that depart to *some* degree from our folk notions of mental representation. In fact, as we will see in chapter 3, notions of representation used in classical computational accounts of cognition are both valuable and somewhat unique to that explanatory framework. What it does mean, however, is that the theoretical notions of representation must overlap sufficiently with our pre-theoretical understanding so that they function in a way that is, indeed, recognizably representational in nature.

When I suggested earlier that our ordinary conception of representation cannot be captured by simple definitions, I did not mean to imply that it can't be illuminated in various ways. If our notion of representation involves a cluster of features, we can ask what some of those features are. In fact, a strong case can be made that there is not one cluster but two overlapping constellations, corresponding with two different families of representational notions. One cluster corresponds to various notions of mental representation, the other to different types of non-mental representation. Cognitive scientists and philosophers often tap into these clusters when they construct theories about the mind that appeal to representations, and as we will see throughout our discussion, the non-mental cluster is often used to explicate cognitive representation. Consequently, it will help to briefly look at some of the aspects of these families of representational notions to get a better sense of where the more scientific notions of cognitive representation come from.

1.1.1 Mental representation within folk psychology

Our ordinary, "folk" conception of mental representation includes things like different types of knowledge, propositional attitudes (beliefs, desires, hopes, etc.), memories, perceptual experiences, ideas, different sorts of sensations, dream states, imaginings, and various emotional responses to circumstances. Some of these notions are clearly closer to what might be considered the "center" of the cluster than others. In particular, our notions of basic *thoughts* – propositional attitudes[2] – appear to be more central to our ordinary understanding of mental representation and most writers treat them as paradigmatic. I'll focus on thoughts in my discussion here (or more accurately, on our conception of thoughts), though a great deal of what I'll say generalizes to other notions of mental representation as well. So, what do we take to be the basic features of thoughts?

It might be supposed that explaining our commonsense perspective on thoughts and other mental representations should be a trivial and uncontroversial affair. Ex hypothesis, our ordinary attitudes about mentality are common knowledge and its main features are easily accessible to all. Alas, things aren't so simple. Exactly what our commonsense understanding of the mind involves and how it works is something heavily debated by both

[2] For those unfamiliar with the term, propositional attitudes are mental states such as beliefs, desires, hopes, fears, assumptions, and the like. They are, as the name implies, a certain attitude (believing, desiring, hoping, etc.) toward a proposition. Propositions are perhaps best conceived of as states of affairs.

philosophers and psychologists; at the present, there doesn't appear to be anything close to an emerging consensus. Since these different accounts of commonsense psychology entail different accounts of how we regard mental representations, it is difficult to articulate this commonsense notion without stepping on someone's toes.

On one side of this debate are many philosophers and psychologists, including myself, who maintain that our commonsense or folk psychology functions as a predictive and explanatory *theory* (Churchland 1981, 1989; Gopnik and Wellman 1992; Stich and Nichols 1993). This view – the "theory-theory" – suggests that, like any theory, commonsense psychology is comprised of both theoretical posits and a number of law-like generalizations. The main posits include various representational states like beliefs, desires and other propositional attitudes, as well as various qualitative states like pains. The "laws" of folk psychology are the platitudes we use to predict and explain one another's behavior. Thus, on most versions of the theory-theory, we treat mental states like beliefs as entering into causal relations that support a wide range of generalizations. One of the more controversial aspects of the theory-theory is that it opens up the possibility of eliminativism – the view that folk psychology might be a radically false theory, and that we will come to discover that its posits, like beliefs and desires, don't actually exist.

However, not everyone accepts the theory-theory account of our ordinary understanding of the mind. Some reject it because they regard belief-desire psychology to be something very different from a system that posits inner causes and law-like generalizations. On one view, it is a way of making sense of the activities of rational and linguistic agents, used to classify and identify rather than to explain and predict. As one author puts it, "[F]olk psychology, so called, is not a body of a theory but an inherited framework of person-involving concepts and generalizations" (Haldane 1993, pp. 272–273). Others reject the theory-theory by claiming that to explain and predict behavior, we rely not on a theory but on a type of simulation. According to this view, we take some of our own information-processing mechanisms "off-line" (so the mechanism generates predictions instead of behavior) and then feed it relevant pretend beliefs and desires that are assumed to be held by the agent in question. Then, sub-consciously, we use our own decision-making mechanisms to generate output which can thereby serve as predictions (and, in other circumstances, explanations) of the agent's behavior. No theoretical posits or laws – just the use of our own machinery to grind out recommended actions that we can then exploit in accounting for the behavior of others (Gordon 1986; Goldman 1992).

Hence, there is considerable disagreement about what our commonsense psychology is really like, which in turn leads to disagreement about what our concepts of mental representation are like. Indeed, there is even disagreement about how we ought to *figure out* what commonsense psychology is really like (Ramsey 1996). So much for the commonality of commonsense! Of course, in presenting our conception of mental representations, there is no way that we can hope to resolve all of these debates here. But for now, given that we just want to get the ball rolling, perhaps we don't need to resolve all of them. Despite the different disputes about the nature of commonsense psychology, there is little disagreement over whether we actually *have* commonsense notions of mental representation. So perhaps there are some basic features associated with those notions that can be agreed upon by most. I think there are at least two.

Intentionality

Most philosophers agree that our concepts of mental representations involve, in some way, intentionality (also referred to as the "meaning," "intentional content," or the "semantic nature" of mental representations). Intentionality (in this context) refers to "*aboutness*."[3] Thoughts, desires, ideas, experiences, etc. all *point to* other things, though they could also, it seems, point to themselves. Intentionality is this feature of pointing, or designating, or being about something. Typically, mental representations are about a variety of types of things, including properties, abstract entities, individuals, relations and states of affairs. My belief that Columbus is the capital of Ohio is about Ohio, its seat of government, the city of Columbus, and the relation between these things. On most accounts, we treat the intentional nature of our thoughts as crucial for their individuation; that is, we distinguish different thoughts at least in part by appealing to what they are about. My belief about the capital of Ohio is clearly a different mental state than my belief about the capital of Indiana. In this way, intentionality serves as a central, distinguishing feature of all mental representations. It is hard to see how something could qualify as a mental representation in the ordinary sense unless it was *about* something – unless it in some way stood for something else.

On most accounts, the intentionality of mental representations is an extremely unique feature of minds and minds alone. While public signs and linguistic symbols are meaningful, their meaning is generally assumed to be derivative, stemming from the conventions and interpretations of

[3] This helpful way of characterizing intentionality is from Dennett and Haugeland (1987).

thinking creatures. That is, the aboutness of a word or road sign is thought to exist only through the aboutness of our thoughts – in particular, the aboutness of the thought that these physical shapes stand for something else. Only thoughts and other mental representations are assumed to have what is called "original" or "intrinsic" intentionality. Intuitively, no one needs to *assign* a meaning to my thought that Columbus is the capital of Ohio for it to be the case that the capital of Ohio is what that thought is about. Such a thought seems to be, as one philosopher has put it, a sort of "unmeant meaner"[4] – a state whose meaning is not derived from other sources. How this is possible is often assumed to be one of the great mysteries associated with mentality.

Along with this "intrinsicality," the intentionality we associate with mental representations brings with it a number of other curious features that have received considerable attention, especially from philosophers of mind. For example, the intentional relation between representation and what it represents is odd in that the latter need not actually exist. For most sorts of relations, both relata are needed for the actual relation to obtain. Yet we can have thoughts about non-existent entities like unicorns and Sherlock Holmes, suggesting the nature of the intentional relation between thoughts and their objects is highly atypical. Furthermore, thoughts can represent the world as being a way that it isn't. Beliefs can be false, perceptual illusions misrepresent reality, and our hopes and desires entertain states of affairs that may never come about. How this is possible is far from obvious. And there is also the curious feature of intentionality referred to as "opacity." Although thoughts are individuated in terms of what they are about, two thoughts about the same state of affairs are not treated as identical. Even though I can be characterized as believing that John Wayne was an actor, I can't be said to believe that Marion Morrison was an actor even though, as it turns out, John Wayne was actually Marion Morrison. The different *ways* we can represent things and events matters a great deal for our ordinary understanding of mental representation.

The oddness of the intentionality we associate with thoughts has led some, most famously Brentano, to suggest that the mind is in some way non-physical. The intentional nature of mental representations is sometimes characterized as an "irreducible" property – a feature that cannot be explained through the natural sciences. Since most contemporary philosophers of mind are physicalists, a major project over the last thirty years has

[4] Dennett 1990. It should be noted that Dennett rejects the idea of intrinsic intentionality and employs the phrase "unmeant meaner" in jest.

been to try to show how we *can*, in fact, explain intentionality in physical terms. Many of these attempts to appeal to the sort of features associated with non-mental representations we will look at in section 1.1.2. While there is considerable debate about how best to explain intentionality, there is near unanimity on the central role it plays in our commonsense understanding of mental representations. Indeed, its importance is so central that there seems to be a tacit assumption held by many philosophers that a theory of intentionality *just is* a theory of representation. As we will see below, this assumption is, for a variety of reasons, highly questionable.

Causality

The second sort of relatively uncontroversial feature associated with mental representations is the set of causal relations that commonsense assigns to our thoughts. Intuitively, mental representations are states that *do* various things. Although philosophers once denied that thoughts could serve as causes (Anscombe 1957; Melden 1961), today there is general agreement that in *some* sense, our ordinary understanding of thoughts attributes to them various causal roles. For example, folk psychology treats my belief that the stove is on as a state with the content *"the stove is on"* employed in a specific range of causal relations. These relations might include being triggered by perceptual stimuli of the dial set to the "on" position, the generation of a fear that my gas bill will be too high, the production of a hand motion that turns the stove off, and so on. On one version of the theory-theory, the set of causal relations associated with our thoughts correspond to the law-like generalizations of our folk psychological theory. A popular example of such a law goes as follows: If someone wants X and holds the belief that the best way to get X is by doing Y, then barring other conflicting wants, that person will do Y. When we explain or predict behavior, the theory-theory claims we (tacitly) replace variables X and Y with whatever propositions we think an individual actually desires and believes. For instance, I might explain Joe's obsequiousness by suggesting that Joe wants a raise and believes the best way to get a raise is by complimenting the boss. This want and belief are together thought to literally cause Joe to act in the way he does. The same applies to other notions of mental representation, including desires, hopes, memories, images, and so on.

While the basic idea that mental representations partake in different causal relations is fairly straightforward and perhaps amenable to scientific treatment, there is, many would argue, a second aspect of our ordinary conception that makes the causal nature of representations more difficult to explain. It has been argued that commonsense psychology suggests that our thoughts not only interact in various ways, but that they participate in these causal relations by virtue of their content (Dretske 1988; Horgan 1989). The type of behavior a belief or desire generates is intuitively determined by what that belief or desire is about. But this makes the causal nature of mental representations more problematic. First, if the causal properties of representations depend upon their intentional properties, then all of the apparent mysteriousness of intentionality extends to their causal nature. Second, many naturalistic accounts of intentional content appeal to head-world relations and historical factors that would seem to have no way of influencing the causal powers of inner cognitive states that might be treated as representations. This has led many to abandon the idea that representations do what they do *because of* what they are about, and instead adopt the weaker position that the causal role of representations *corresponds* to their content in such a way that their interactions "make sense." If I believe that if P then Q and also come to believe P, then these two beliefs will cause me to believe Q, though not, strictly speaking, by virtue of their content (which is causally inert). In the next chapter, we will look at how this story is presented in the framework of the CCTC, while in chapter 4 we'll examine a theory that attempts to show how content *is* causally relevant.

Beyond these mundane observations about the intentionality and causality of mental representations, what little consensus there is about our commonsense picture of mentality begins to evaporate. For example, the relevance of other factors for our basic conception of mental representation, such as the role of consciousness, public language, or rationality, is far more controversial.[5] Still, it might be thought that from this very modest analysis, we have enough to begin to see what a psychological theory that appeals to inner representations ought to look like. Mental representations are states that have some sort of non-derived intentionality and that interact with other cognitive states in specific sorts of ways. Since folk psychology is arguably a primary source from which a representational perspective is derived, one could say its posits should be all that a scientific theory needs to invoke. A psychological theory that invokes inner representations is thereby a theory that invokes beliefs, desires, ideas, and other folksy notions.

[5] In fact, things are more controversial than I've even suggested here. For example, as noted above, Daniel Dennett denies that there is such a thing as original intentionality. He also rejects the idea that we treat mental representations as causes in any straightforward, billiard-ball way (Dennett 1991).

While it is true that our commonsense notions of mental representation influence psychological theorizing a great deal, it would be a mistake to assume that cognitive scientists set out simply to mimic these notions when developing their own picture of how the mind works. As we'll see in the coming chapters, researchers develop and produce theoretical notions of representation that depart in various ways from folk notions. Even when notions like belief are incorporated into scientific accounts, it is typically stretched and modified in order to fit the explanatory needs of the theory. Moreover, our ordinary notion of mental representation leaves unexplained a great deal of what a theory-builder should explain about how something actually serves as a representation. Commonsense psychology provides us with little more than a crude outline of mental representations and leaves unanswered several important questions about how representations drive cognition. This point will be addressed in greater detail in section 1.2 below. But before we examine that topic, we should also briefly consider our ordinary notions of non-mental representation.

1.1.2 Non-mental representation

As with mental representation, the commonsense class of non-mental representations is quite large and encompasses a diverse range of states and entities. These include, but are not limited to, linguistic symbols, pictures, drawings, maps, books, religious icons, traffic signals and signs, tree rings, compass needle positions, tracks in the snow, hand signals, flashing lights, and on and on. This diversity suggests that whatever non-mental representation amounts to, there are few restrictions on the types of things that qualify. Perhaps this is unsurprising if, as suggested earlier, non-mental representations all have derived intentionality. If something's status as a representation is merely assigned by minds, and if minds can assign meaning to practically anything, then we would expect there to be a very diverse array of things that serve as non-mental representations. Moreover, if non-mental representation is entirely dependent upon mental representation, it is far from clear that there is much that the former can tell us about the latter. If non-mental representations lack the central defining features associated with cognitive representations, why should we bother thinking about non-mental representations at all?

There are a couple of answers to this question. First, some have argued that it is just wrong to suppose that only mental states possess intrinsic intentionality. They have suggested that there is a type of low-level meaning "out there" in the world, possessed by physical states without the intervention of interpreting minds. For example, some authors have claimed that a tree's rings carry information about its age all by themselves, irrespective or whether or not anyone notices this (Dretske 1988). If this is correct, then it may be possible to gain some sort of insight into cognitive representation by exploring the representational character of things that are non-mental. Second, even if all non-mental representation is in some sense derivative, we might still be able to learn important facts about the nature of representation – especially about the way cognitive scientists *think about* representation – by looking at the non-mental cases. Since we are trying to gain some insight into the sort of thing researchers have in mind when they posit representations in psychological theories, it is worth at least considering the type of representations we encounter in our everyday lives.

Historically, there have been many attempts to spell out the central features of everyday representations. One such attempt is Charles Peirce's theory of semiotics (1931–58), an extremely rich but also cryptic analysis of the general nature of representation. Despite the abstruse nature of Peirce's theory, it provides at least a basic framework from which helpful insights about the character of non-mental representation can be found.[6] Since I plan to pillage the parts of Peirce's account that I find intuitively plausible, my apologies to Peirce scholars for ignoring or mistreating various nuances of his view.

One of Peirce's main contributions on representation is an analysis of the different ways in which representations – what he calls "signs" – are linked to the things they represent. Peirce appeals to three types of content "grounding"[7] relations, corresponding with three different sorts of signs. First, there are "icons," signs that are connected to their object by virtue of some sort of structural similarity or isomorphism between the representation and its object. Pictures, maps, and diagrams are all iconic representations. A picture represents a person at least in part because the former closely resembles the latter. Second, there are "indices" – signs that designate things or conditions by virtue of some sort of causal or law-like relation between the two. An array of tree rings exemplifies the category of indices since the age of the tree reliably causes the number of rings. What many philosophers today would call "natural signs" or "indicators" qualify as Peirce's indices. Pierce's third category is what he calls "symbols."

[6] See, for example, Barbara von Eckardt's (1993) excellent synopsis of Peirce's account that refines and modifies his view, highlighting its most salient and plausible components for cognitive science.

[7] Like others, I use the phrase "content grounding relation" here to designate the natural conditions or relations that are thought to give rise to the intentional relation between a representation and its object.

Symbols are connected to their objects entirely by convention. There is no further feature of a symbolic sign that bestows their content – they mean what they do entirely by stipulation. Linguistic tokens, such as written words, are paradigm cases of Peirce's symbols.

Peirce's analysis is important for our purposes because, as it turns out, these same ideas serve as the basis for different notions of representation found in cognitive science. In fact, much of what has been written about mental representation over the last thirty years can be viewed as an elaboration on Pierce's notions of icons, indices and symbols. In chapter 3, we'll look at how something quite similar to Pierce's notion of representational icons is actually an important theoretical entity in the accounts of cognition put forth in the CCTC. As we'll see, many versions of the CCTC posit inner states that serve as representations in the same sense in which the lines and figures on a map serve to represent features of some terrain. In chapter 4, we'll look at notions of representation that are based on the same sort of law-like dependencies Pierce associated with indices that appear in many of the newer accounts of cognition. In both sorts of cases, a notion of representation is put forth in accounts of cognitive processes that is based upon principles associated with our pre-theoretical understanding of non-mental representation. Our job will be to determine whether these principles can be successfully applied to cognitive states and processes in the brain so that we wind up with an explanatorily adequate account of cognitive representation.

While philosophers and cognitive scientists have attempted to explain representation by appealing to the physical relations associated with icons and indices, Peirce himself would probably regard this whole project wrong-headed. Peirce held that representation is always a triadic relation, involving (a) the sign, (b) its object, and, (c) some cognitive state of an interpreter (the "interpretant"). On Peirce's account, the interpretant is itself a representation, leading to a regress which he cheerfully embraces. But here we can simply treat the third element as a cognitive agent. For Peirce, all three elements must be involved; if any one component is missing, there is no representation. Consequently, the Peircean picture rejects any attempt to reduce representation to a dyadic relation that excludes the interpreter. For him, there can be no meaning or representational content unless there is some thing or someone *for whom* the sign is meaningful.

This makes it sound as though Peirce was a strong advocate of the original/derived intentionality distinction that, as we suggested at the outset of this section, threatens to undercut any attempt to explain mental representation in terms of what we know about non-mental representations. But it is somewhat doubtful that Peirce had in mind the original/derived intentionality distinction, since for him even mental representations have, in some sense, derived intentionality. For Peirce, *all* forms of representation involve something like an interpreter, and it is far from clear that he distinguished mental and non-mental representations in the way many do today. What *is* significant about Peirce's triadic analysis is the idea that representations are things that are *used in a certain way.* Something qualifies as a representation by playing a certain kind of role. Similarly, Peirce treats representation as a functional kind – to be a representation is to be something that does a certain job (Delaney 1993, pp. 130–156).

Peirce seems right that this is a basic feature of our ordinary understanding of representation. As Haugeland puts it, "representing is a functional *status* or *role* of a certain sort, and to be a representation is to have that status or role" (Haugeland 1991, p. 69). When we consider non-mental representations like maps, road signs, thermometers and bits of language, it is clear that these things are employed by cognitive agents as a type of tool. These are all things that serve to inform minds in some way or other about various conditions and states of affairs, and outside of that role their status as representations disappears. The proverbial driftwood washed up on an uninhabited beach does not, intuitively, represent anything, even if it spells out "UNINHABITED BEACH" or is arranged in a way that maps a course to a nearby lake. However, if someone were to come along and use the driftwood as a type of map, then it would indeed take on a representational role.

What all of this suggests is that if our understanding of cognitive representations is based on our understanding of non-mental representations, then we need to understand how something can play a representational role *inside* a given cognitive system. If our basic notion of non-mental representation is a functional notion, like our notion of a hammer or door stop, then any cognitive theory positing inner states derived from these notions is positing states that have a job to perform. With non-mental representation, that job appears to require a full-blown cognitive agent as an employer. Exactly what that job is supposed to entail *within* a cognitive agent – in the context of psychological processes – is far from clear. If ordinary notions of non-mental representation are to form the basis for understanding representation in cognitive theories, and if those ordinary notions always presuppose some sort of representation *user*, then we need to provide some sort of account of representation use where the user isn't a full-blown mind. In the following chapters, I'll argue that we can do this for one of Peirce's signs – namely, icons – but not the others. I'll argue that things that represent in the way icons represent can be found within a mechanical or biological system, whereas this can't be done for things that represent in the manner of Peirce's symbols or indices.

While the preceding is hardly an exhaustive analysis of the nature of non-mental representation, it has highlighted two important ideas. The first is that there are basic kinds of non-mental representation and that these are also found in theories of how the mind works. Hence, theorists appeal to certain sorts of non-mental representation – discussed by Peirce – as a guide for understanding the nature of cognitive representation. The second point is that there are some *prima facie* difficulties associated with such an appeal. Central among these is the fact that our ordinary conception of non-mental representation seems to pre-suppose that they do a certain kind of job (like informing or designating), and it is far from clear how we are supposed to characterize that job without appealing to an up and running mind. Because I think this last point is extremely important (and often underappreciated) it will help to consider it more closely.

1.2 THE JOB DESCRIPTION CHALLENGE

In the last section, we were operating on the assumption that by reflecting a bit on our ordinary notions of representation, we could gain a better understanding of what it is that scientists are referring to when they claim the brain uses such states. But one might wonder why we need to look at commonsense notions of representation at all. In most scientific theories, a theoretical notion is introduced in such a manner that includes the unique properties that give the posit its specific explanatory role. The positing of genes, for example, involves a specification of the different relations and causal roles that describes the sort of job we think genes perform (Kim 1998). By using this job description, we can then go look for the actual bio-chemical structures that fit the bill. Of course, along the way we may discover that our job description needs to be modified in some way. But we cannot make any progress in understanding how a given posit is actually realized unless we first have a fairly clear understanding of what it is the posit supposedly does. This understanding is not provided by commonsense, but by the scientific theory itself.

In the case of representation, however, things are more complicated. As we've just seen, representational notions already have a home in our non-scientific conception of the world. This non-theoretical understanding constrains the sorts of things that can qualify as representational states, even in the context of a scientific theory. As we noted in section 1.1, the scientific notions must in *some* way be rooted in our ordinary conception of representation; otherwise, there would be little point in calling a neural or computational state a representation. Thus, we briefly looked at two sets of commonsense notions of representation to try to get a clearer sense of exactly what is being invoked when a theorist posits inner representations as an element of the mind. What we would like these notions to provide is a specification of the essential or core features of representation that we can then use in our assessment of scientific theories that claim to be representational. We would like something akin to a *job description* for representational posits that is analogous to what we have for other theoretical posits, like genes or protons, so that we can then determine if a given state or structure fits the bill.

Yet as we saw in the last section, an analysis of the commonsense notions doesn't really provide us with what we are after. The problem is *not* that the commonsense notions don't involve core features or offer job descriptions for representational states. Rather, the problem stems from the sort of features and roles that are associated with these notions. In the physical or biological sciences, a job description for a posit can be provided in straightforward causal/physical (or causal/bio-chemical) terms, like the pumping of ions or the production of some enzyme. But in the case of both non-mental and mental representation, the relevant roles include things like *informing*, *denoting* or *standing for something else*. It is not at all clear how these sorts of roles are supposed to be cashed out in the naturalistic, mechanistic[8] framework of a cognitive theory. Many scientific theories of the mind attempt to explain cognition in neurological or computational terms. But our ordinary understanding of representation involves features and roles that can't be translated into such terms in any obvious way.

Consider our ordinary notions of mental representation. As we saw above, our commonsense understanding of beliefs, desires, and other folk representational states assigns to them some sort of underived or intrinsic intentionality, and this feature is thought to be central to their serving as representational states. But intentionality clearly isn't a basic causal or functional property. Consequently, when we look inside a physical

[8] By "mechanistic," I simply mean an explanation that appeals to physical or perhaps what are often called "syntactic" states and processes. A mechanistic explanation accounts for some capacity or aspect of the mind by showing how it comes about through (or is realized by) structures, states and operations that could be implemented in physical processes.

system to determine if there are mental representations, it is not at all clear what we are looking for. It isn't clear what having the property of "aboutness" is supposed to entail for a state of a physical system, or how having such a feature will influence the way a physical system operates. If researchers simply adopt, without further elaboration, our ordinary notions of mental representation as part of their naturalistic accounts of the mind, we are left with an account that can't be fully understood because we have no sense of what serving as a representation in such a system is supposed to entail.

A similar problem arises with regard to our commonsense understanding of non-mental representation. Recall that here the notion of representation is associated with a user, and if we ask about the sorts of things that use such representations, the most natural answer would be a full-blown cognitive agent. Everyday examples of non-mental representations – road signs, pieces of written text, warning signals, and so on – all involve thinking agents who use the representation to stand for something else. How, then, can we specify the functional role of representation as something employed *within* cognitive systems, when it intuitively functions as something used externally *by* cognitive systems? As Dennett points out,

> nothing is intrinsically a representation of anything; something is a representation only for or to someone; any representation or system of representation thus requires at least one user or interpreter of the representation who is external to it. Any such interpreter must have a variety of psychological or intentional traits . . .: it must be capable of a variety of comprehension, and must have beliefs and goals (so it can use the representation to inform itself and thus assist it in achieving its goals). Such an interpreter is then a sort of homunculus . . . Therefore, psychology without homunculi is impossible. But psychology with homunculi is doomed to circularity or infinite regress, so psychology is impossible. (1978, p. 122)[9]

What all of this suggests is the following. If cognitive scientists are going to invoke a notion of representation in their theories of cognition, then although such a posit will need to share some features with our commonsense notions (to be recognizable as representations), the scientific account cannot simply transplant the commonsense notions and leave it at that. The folk notions, as such, are ill-suited for scientific theories because they carry features whose place in the natural order is unspecified. Hence, some further work is needed to account for these features and show how

[9] Dennett argues this dilemma is solved in the CCTC through the "discharging" of the interpreter/homunculus. This involves explaining the sophisticated capacities of the homuculus/interpreter by appealing to increasingly less sophisticated components that comprise it. We will again return to a discussion of this strategy in forthcoming chapters.

representation can be part of a naturalistic, mechanistic explanation. There needs to be some unique role or set of causal relations that warrants our saying some structure or state serves a representational function. These roles and relations should enable us to distinguish the representational from the non-representational and should provide us with conditions that delineate the sort of job representations perform, *qua* representations, in a physical system. I'll refer to the task of specifying such a role as the "*job description challenge.*" What we want is a job description that tells us what it is for something to function as a representation in a physical system.

What might a successful job description for cognitive representation look like? Part of this will depend on the particular sort of representational notion invoked in a given account. But there are more general criteria that we can expect to be eventually met whenever a notion of inner representation is put forth as part of a naturalistic theory of cognition. These are conditions that need to be elucidated if the invoking of inner representations is going to do any real explanatory work. In the case of reductive theories – that is, theories that attempt to explain cognition as something else (like computation or neurological processes) – representation cannot simply serve as an explanatory primitive. If we are to understand these processes as representational in nature, we need to be told, in presumably computational, mechanical or causal/physical terms, just how the system employs representational structures. Principally, there needs to be some sort of account of just how the structure's possession of intentional content is (in some way) relevant to what it does in the cognitive system. After all, to be a representation, a state or structure must not only have content, but it must also be the case that this content is in some way pertinent to how it is used. We need, in other words, an account of how it actually *serves as* a representation in a physical system; of how it functions as a representation. Dretske captures exactly the right idea: "The fact that [representations] have a content, the fact that they have a *semantic* character, must be relevant to the kind of effects they produce" (Dretske 1988, p. 80). For the moment, we can leave unspecified exactly what "relevant" means in this context. As we saw in section 1.1.1, specifying the relevancy is tricky business because on several accounts of content, it is far from clear how the content itself can be a *causally* relevant feature of a structure. And if these conditions aren't causally relevant, it is far from clear how they can be explanatorily relevant or how they can be at all "relevant to the kind of effects they produce." For now, we can simply stipulate that the positing of inner representations needs to include some sort of story about how the structure or state in question actually plays a representational role.

Specifying how a posited representation actually serves as a representation is important because representation is, as Pierce and others have emphasized, a functional notion. Without the functional story, it would be virtually impossible to make sense of this aspect of the theory. Consider the following analogy. Suppose someone offers an account of some organic process, and suppose this account posits the existence of a structure that is characterized as a pump. The advocate of the account would need to provide some sort of explanation of how the structure in question actually serves as a pump in the process in question. Without such a story, we would have no reason for thinking that the description is accurate or that there are any structures that actually *are* pumps. Now suppose that when we ask how it is that the structure in question functions as a pump, we are told that it does so by absorbing some chemical compound, and nothing more. In this scenario, we would properly complain that the role the structure is characterized as playing is not the role associated with our ordinary understanding of a pump. To be a pump, an entity must, in some way, transfer material from one place to another. What the theory appears to posit is not a pump, but instead what sounds more like a sponge. Because functioning as a sponge is clearly different than functioning as a pump, then despite the way the theory is advertised, it would belong in the class of sponge-invoking theories, not pump-invoking theories.

In a similar manner, cognitive researchers who invoke a notion of inner representation in their reductive accounts of cognition must provide us with some explanation of how the thing they are positing actually serves as a representation in the system in question. We need to be given a description of the structure that enables us to see how it does something recognizably representational in nature. If we are told that it is a representation by virtue of doing something that no one would think of as a representational role – say, by functioning as a mere causal relay – then we would have good reason to be skeptical about the representational nature of the account. Indeed, if the role described is one that is shared by a wide array of other types of entities and structures, we would have the additional problem of representations (in the dubious sense) appearing everywhere. In other works, I've referred to this as the "problem of pan-representationalism" (Ramsey 1995).[10] A central goal of this book is to argue that this hypothetical situation is in fact the actual situation in a wide range of newer cognitive theories.

[10] Fodor has used the term "pansemanticism" to make a similar point (Fodor 1990).

It is important to recognize that meeting the job description challenge is not the same thing as providing a naturalistic account of content. The latter would present the set of physical or causal conditions that ground the content of the representation – the conditions that determine how a state or structure comes to have intentional content in the first place. A complete and fully naturalistic account of representation would need to provide such an account since without it a central aspect of representation would remain mysterious and unexplained. As Dennett might put it, the "intentionality loan" associated with the positing of a representation would go unpaid (Dennett 1978). As noted above, providing such a set of conditions has been a major project in the philosophy of mind for some time. Some of the more popular attempts to explain content naturalistically are accounts that appeal to types of nomic dependency relations (Dretske 1988, Fodor 1987), causal links to the world (Field 1978), evolutionary function (Millikan 1984), and conceptual roles within a given system (Block 1986). Yet insofar as the goal of these theories is to explain a certain type of *relation* – the relation between a representation and its intentional object – they are not accounts that directly explain what it is for a state or structure to actually *function as* a representation in a physical system. As we will see in forthcoming chapters, it is true that some of the accounts of content also strongly suggest certain strategies for answering the job description challenge. But viewed strictly as accounts of content, that is not their primary objective.

To see this distinction better, consider the various circumstances associated with the normal use of a compass. On the one hand, we might be interested in knowing how a compass actually functions as a representational device. What makes a compass serve as a representational device is the fact that the position of the needle literally serves to inform a cognitive agent about different directions. The content of the compass is relevant to its job because the needle's position is used to reveal facts about, say, the orientation of magnetic north. It is in this way that the compass comes to serve as a representation. There are, of course, many ways it might do this, and thus there are many different types of compasses. For example, a pocket version of a compass, which is simply held in the hand needle-side up, operates in a manner very different from the stationary versions, like those permanently mounted on the dash of vehicle (which may not even use a needle). But with all compasses, we can see how they play a representational role by seeing how they serve to inform people about directional orientation. This is an understanding of the functionality of a compass.

On the other hand, we might instead be interested in knowing what conditions are responsible for the representational content of the compass – something relevant to, but very different from, the compass's functional role. If we wanted to know how a compass came to acquire its intentional content – the conditions that underlie the information it provides – one obvious answer would be to say that the needle of the compass comes to designate magnetic north because its position is nomically dependent upon (or reliably co-varies with) magnetic north. The semantic content of the compass is thereby grounded in a dependency relation between the needle and a certain condition of the world. It is this dependency relation that makes the needle's position entail certain facts about the world, and thereby enables us to use it as a representational device.

The point here is that the account of how the compass serves as a representation is different from the account of the conditions responsible for its representational content. Of course, in one clear sense, the compass is a poor example for our purposes because it serves as a representation only for a full-blown interpreting thinker, and a thinker is the very sort of thing a theory of the mind can't invoke. But the compass illustrates the need to bear in mind that understanding how a state's content is grounded in some set of conditions is not the same thing as understanding how the state actually serves as a representation. Someone could perfectly well understand the nature of the dependency between the needle and magnetic north and yet be completely ignorant about the way in which the compass functions as a representational device. This bears emphasis because writers sometimes treat naturalistic theories of content as though they provide a complete account of cognitive representation. However, a theory of content is only one part of the puzzle. Any theory that invokes representational structures should also include an accounting of how the posit functions as a representation. The latter would include some sort of accounting of how something's status *as* a representation is pertinent to the way the cognitive system performs some cognitive task.

In fact, in cognitive research, the need to answer the job description challenge for a representational posit is far more pressing than the need to provide a naturalistic account of content. To some extent, researchers can leave the explanation of content to the philosophers. If theorists can develop accounts of cognitive processes that posit representations in a way that reveals just how the posit plays a representational role in the system, then the explanation of content can wait. They can say, "Look, I'm not completely sure how state X comes to get the content it has, but in my explanation of cognition, there needs to be a state X that serves as a representation in the following way." So from the standpoint of psychological theory development, the need for an account of content-grounding is not so urgent. However, if a theorist cannot explain the sense in which a representational posit actually *serves as* a representation, or offers an explanation that is grossly inadequate, then the very representational character of the theory is seriously undermined. In fact, we would have no real reason to think the account is actually representational at all, and whatever pre-theoretic understanding of representation we possessed would be irrelevant to our understanding of the cognitive account on offer. In short, a representational theory of cognition should provide, at a bare minimum, an explanation of how something serves as a representation in such a way that, at the end of the day, we still have a *representational* theory, instead of a non-representational account of a psychological process.[11]

The crux of the job description challenge, then, is one of steering a course between the Scylla of putting forth conceptions of representation that are too strong (because central aspects of representation are left unexplained) and the Charybdis of positing conceptions of representation that are too weak (because representation is reduced to something non-representational, uninteresting and ubiquitous). In the case of the former, we would simply have the reintroduction of a sophisticated cognitive capacity (the use and interpretation of representations) with no real understanding of how this is done. In the case of the latter, we would have structures that operate in a way that *is* intelligible, but not intelligible as playing a representation role. What we want is an account of how something described as a representation functions *as such* in a computational or biological system. We want an account that allows us to intuitively recognize the processes in question as distinctively representational, and at the same time illuminates how this comes about.

While I think many authors have recognized certain aspects of the job description challenge, the full nature of the challenge has not been adequately appreciated either by researchers who invoke representations in their theories, or by philosophers attempting to explain representation. It might be assumed that those writers who develop "teleo-semantic" accounts of content come the closest to addressing the worry because their accounts are built on the idea that an appeal to proper function is

[11] In some accounts of higher order representational phenomena, such as memory or conceptual representation, "lower order" representations of features are invoked to serve as constituent elements of larger representational structures. In one sense, these feature representations are often introduced by mere stipulation, without any account of what they are doing that *makes* them into representations of features. Yet at the same time, it could be argued that they clearly are serving as representations in the proposed architecture by functioning as representational constituents of some larger representational system; a fuller discussion of this sort of representation is offered in chapter 3.

the key to understanding cognitive representation. Yet because many of these accounts are focused on handling worries associated with the naturalization of the content relation, the functional role of representing is often invoked without being fully addressed. That is, we are often told that a structure's proper functioning as a representation is critical for understanding, say, how the structure can misrepresent, without being told in sufficient detail what proper functioning as a representation amounts to. For example, Millikan (1984) attempts to provide a general framework that applies to both mental *and* non-mental representations. On one reading of her account, to function as a representation is to be "consumed" by an interpreter that treats the state in question as indicating some condition. Following Peirce, this seems reasonable for non-mental representations – cognitive agents, including non-humans, take advantage of signs and signals in various ways. But insofar as the same account is supposed to apply to internal cognitive representations, it is far from clear how we are supposed to make sense of representation consumption *inside* a cognitive system. Even worse, Millikan seems to allow that processes normally assumed to have a non-representational function, such as the flow of adrenalin caused by threatening situations, really are quasi-representational after all (Millikan 1993). Ultimately, just how a state or structure actually serves a distinctive role of representing is left somewhat mysterious on Millikan's account, and she thus fails to directly answer the job-description challenge. Millikan's account of representation is unsatisfying because she leaves the functionality of representation, in a sense, "under-reduced."[12]

Dretske (1988), on the other hand, does provide an explication of what it is for something to function as a representation in purely mechanistic terms. On my view, Dretske raises exactly the right questions and addresses exactly the right issues. But as we will soon see, it is hard to understand why a structure functioning in the manner Dretske describes should be seen as representational at all. Dretske's account runs into trouble because representation is, in a sense, "over-reduced" – that is, it is reduced to a set of conditions and relations that intuitively have nothing to do with representation at all. In chapter 4, this critique of Dretske's theory will be spelled out in more detail.

It might seem that as I've explained things here, the job description challenge is quite literally impossible to answer. Either you reduce

[12] In fairness to Millikan, it should be noted that her account is quite complex and lends itself to different readings, one of which suggests a notion of representation that *does* meet the job description challenge in the manner that will be suggested in chapter 3.

representation to causal–physical conditions or you don't. If you do, then I'll say you have reduced representation to the non-representational, and therefore you have abandoned the notion of representation altogether. If you don't do this, then you've left some aspect of representation unexplained and mysterious. So, it seems, you're damned if you do and damned if you don't. However, I think it is not only possible to meet the job description challenge, but that this has been successfully done with certain theories in the CCTC paradigm. Chapter 3 will be devoted to spelling out exactly how this works. The point here is that some ways of fitting representations into the natural order reveal why it makes sense to view a given structure as a representation, whereas other ways fail to do this. My claim is that the difference between the two roughly corresponds to the division between classical computational theories and the newer, non-classical accounts of cognition.

Before moving on, it is important to be very clear on exactly what answering the job description challenge involves. The challenge involves answering neither of the following questions:

(a) Is it possible to describe physical or computational processes in representational terms?
(b) Is it absolutely necessary to describe physical or computational processes in representational terms?

The shared problem with these two questions is that they lend themselves to trivial answers that are uninformative. Consider question (a). As Dennett (1978) and others have noted, it is indeed possible to adopt the "intentional stance" toward practically every physical thing and system. Even a rock can be described as acting on the belief that it needs to sit very still. So, trivially, it is always *possible* to characterize physical systems using representational language by adopting this perspective. However, we tend to find this sort of intentional characterization gratuitous and unnecessary, in part because the notion of representation involved fails to be sufficiently robust. Showing that a system is representational in *this* sense isn't terribly informative in helping us to understand what might be going on with representational cognitive systems.[13]

Going the other way, it is always possible to describe a physical system using purely non-intentional, causal/physical terms. Just as we can avoid biological language in the description of biological systems by dropping

[13] Dennett himself would disagree, as he believes it is a mistake to try to characterize mental representation as a concrete state playing a certain role. Instead, he believes the distinction between representational and non-representational systems is entirely a function of the usefulness of adopting the intentional stance. For Dennett, the whole enterprise of trying to understand representation in the manner described here is misguided.

down to the level of molecules and atoms, so too, it will never be absolutely *necessary* to invoke representational language in the characterization of a representational system. So the answer to (b) is trivially "No." Invoking Dennett's terminology once again, it is always in principle possible to adopt the "physical stance" (using only the terms of basic physics) toward any physical system, even when robust and sophisticated representations are working within the system.

Hence, the job description challenge requires us to address questions that are more nuanced than these. The sorts of questions that need answering are more along the lines of the following:

(c) Is there some explanatory benefit in describing an internal element of a physical or computational process in representational terms.
Or maybe:

(d) Is there an element of a proposed process or architecture that is functioning as a representation in a sufficiently robust or recognizable manner, and if so, how does it do this?
Or even:

(e) Given that theory X invokes internal representations in its account of process Y, are the internal states actually playing this sort of role, and if so, how?

Now unfortunately, neither (c) nor (d) nor (e) is as clean or crisp as we would like. Exactly what counts as representations having an "explanatory benefit" in a theory, or just what it means to be a "sufficiently robust" notion of representation – these are vague matters that, to some degree, require our making a judgment call. But as noted earlier, this shouldn't be terribly surprising. We want to know if, given what a theory says, something is actually functioning in a manner that is properly described as representational in nature. We can't do this without making use of our ordinary understanding of representation and representation function. And, as with any form of concept application, this requires a judgment call. So be it. As we proceed with our analysis, the outlines of what is and is not involved in a "sufficiently robust" notion of representation having an "explanatory benefit" will start becoming increasingly clear.

1.3 DEMARCATING TYPES OF REPRESENTATION AND TYPES OF REPRESENTATIONAL THEORIES

Since we are going to explore different notions of representation in cognitive science with regard to how well they meet the job description challenge, more needs to said about the demarcation strategy I plan to use to classify different notions of representation. There are, of course, a variety of different ways we can distinguish representational notions. The most popular strategies appeal to either the way in which the representation is thought to acquire its content (e.g., nomic dependency vs. conceptual role), or the form or structure of the representation (e.g., compositional form vs. distributed representation), or the sort of theory in which it appears (the classical computational vs. connectionist models).[14]

While all of these taxonomies have their advantages, I am going to use a scheme that is better suited for the issues we will be addressing. In what follows, we need to demarcate notions in a way that places the explanatory role of representations at center stage. My taxonomy will individuate types of representations in terms of the conditions thought to be constitutive of the representational role. That is, I'll be grouping together representational posits from various theories if they all appeal to the same or similar factors to justify the claim that something is serving as a representation. For example, one representational notion we will explore is based on the idea that structures function as elements of a model or simulation of some target domain. Another notion stems from the idea that something functions as a representation because it is reliably triggered in a certain way. Carving things up this way will do two things for us. First, we can avoid examining each individual theory in cognitive science because we can cluster theories together that employ the same basic representational ideas. Second, it will focus attention on what I take to be the most critical feature of representational posits. By making the criteria for type-identity those factors in virtue of which something is claimed to serve as a representation, we will direct the spotlight on the aspect that matters the most for our interests.

Another issue that needs clarifying concerns the way representation is linked to the overarching explanatory goals of a cognitive theory. I've been treating notions of representation as a type of theoretical posit, put forth as part of an explanatory apparatus to account for some cognitive ability or process. On this construal, notions of representations are *explanantia*. However, a great deal of work in cognitive science is also devoted to explaining the nature of representation itself. Various accounts of knowledge, memory, imagery, concepts and other intentional aspects of the mind take representation to be the very cognitive phenomenon that one is attempting to explain. On this construal, cognitive representations are the *explanandum* of a given theory. How does my analysis bear on theories that don't posit representational states but rather try to explain them?

[14] See, for example, Fodor (1985).

First, the distinction between theories that posit representations and theories that try to explain representation is not as sharp as one might assume. A large number of cognitive models – indeed, perhaps the majority – do a little of both. For example, production systems like Allen Newell's SOAR model can be seen as an attempt to explain how humans do certain types of problem solving by providing an account of a specific style of knowledge representation and retrieval (Newell 1990). Newell's theory employs a computational framework to explain both general cognition and the way we represent problems and their solutions. Moreover, we've seen that reductive theories that posit representations (as explanantia) also offer a story about the way representations are implemented in the structures or states of their overarching architecture. This part of the account works as a mini-theory about the nature of representation itself. Thus, the division between theories that treat representation as explanantia and those that treat them as explananda is not very sharp.

Second, it seems clear that meeting the job description challenge should be a goal for theories of representation every bit as much as it is for theories of cognition that invoke representations. Surely any reductive account that is designed to explain the nature of mental representation, or a specific sort of mental representation, will need to spell out how the system employs the state or structure *as* a representation. For example, if the theory is a reductive account of memory that attempts to explain how we store and retrieve representations of long-term knowledge, then it will need to be shown how the parts of the system allegedly responsible for this really do function as representations with intentional content. Answering the job description challenge should be a primary goal of any theory of representation. Hence, much of what follows will be relevant to those reductive accounts whose primary goal is to explain cognitive capacities by appealing to representations, *and* those whose main objective is to explain mental representation itself.

1.4 SUMMARY

In this chapter, the aim has been to introduce some of the issues and concerns that will occupy us in the subsequent chapters. We've seen some of the intuitive aspects of our commonsense understanding of representation that can and do become incorporated into the more theoretical notions found in cognitive science. We've also seen some of the problems associated with those intuitive aspects. One problem is determining how representational content fits into the natural world. But cognitive theories that invoke representations carry the greater burden of providing an adequate account not of content, but of representational function – of how something serves as a representation in the proposed architecture. This is the job description challenge, and we will be returning to it throughout our analysis of different conceptions of representation. The challenge is to explain how a given state actually serves as a representation in a way that is both naturalistically respectable and doesn't make representation completely uninteresting and divorced from our ordinary understanding of what representations are.

In what follows, we will see that some theories employ notions of representation that lend themselves to a successful assimilation into the natural order while retaining their status as real representations. Other theories employ notions of representation that do not successfully assimilate. By and large, the successful notions are found in the CCTC while the unsuccessful notions can be found in connectionist models or the various accounts in the cognitive neurosciences. So I'll be offering both a positive and a negative account of representation in the cognitive sciences. However, before we can start comparing these different accounts, we first need to remove what I take to be an ill-conceived interpretation of the notion of representation in the CCTC. We saw above that naturalistic theories cannot just co-opt the folk notions of mental representation when constructing their accounts of cognition because the folk notions come with features whose place in the natural order is unclear. Further explication is required. But another danger is that the intuitive nature of our commonsense framework will permeate and cloud our understanding of what might actually be a more technical notion of representation that is not directly based on folk psychology. In large measure, I believe this has happened with our current understanding of the CCTC, resulting in a mistaken interpretation of how the CCTC is committed to inner representations. Showing this will be the goal of the next chapter.

2

Representation in classical computational theories: the Standard Interpretation and its problems

In this chapter I reveal what I take to be a popular set of assumptions and tacit attitudes about the explanatory role of representation in the CCTC. I'll suggest that these assumptions and attitudes collectively give rise to an outlook on representation that amounts to a sort of merger between classical computational theory and folk psychology. In other words, the way researchers and especially philosophers have come to regard the importance of representations in the CCTC has been largely determined by their understanding of beliefs and other commonsense notions. This has led to a way of thinking about computational representations that suggests their primary explanatory function is to provide a scientific home for folk notions of mental representations. In other words, symbolic representations in the CCTC have come to be viewed as the scientific analogues for beliefs, desires, ideas, thoughts, and similar representational posits of folk psychology.

This perceived connection between computational representation and folk psychology comprises what I will refer to as the "Standard Interpretation" of the CCTC. While few authors have explicitly stated the Standard Interpretation, at least not in the manner in which it will be presented here, it (or some version of it) has nevertheless played a significant role in shaping the way many people think about the CCTC. After spelling out what I think the Standard Interpretation involves, I'll try to show that it leads us down a path where, despite various claims to the contrary, we wind up wondering whether the symbols of classical models should be viewed as representations at all. This path has been illuminated by two important skeptics of CCTC representationalism, John Searle and Stephen Stich. Searle and Stich both exploit the alleged link between the CCTC and folk psychology to challenge the claim that the CCTC can or should appeal to inner representations. Searle does this by denying that classical symbols can capture what is essential about thoughts, while Stich does this by arguing that beliefs and other folk notions don't belong in a serious scientific account of cognition. Both criticisms can be seen as ways of making the more general point that if classical symbols are to serve as *reducers* for folk notions, the job description challenge goes largely unanswered.

My own view is that the Standard Interpretation has clouded our understanding of the explanatory work that is *actually* done by the notion of representation in classical computational theory. The mistake is in thinking that computational symbols need to capture ordinary folk psychological notions in order to qualify as representational states. In the next chapter, I'll argue that these problems disappear once we abandon the Standard Interpretation and recognize that the sort of representations invoked by the CCTC are not based directly on folk psychology, but on the kinds of explanatory strategies used by the CCTC. For now, however, my aim is only to present the Standard Interpretation and show how its defense of representationalism runs afoul. To show this, I'll first present some of the basic principles behind the CCTC. Then I'll show how on the Standard Interpretation, computational symbols come to be linked with notions like belief. Finally, I'll present Searle's and Stich's criticism of representationalism, and examine the ways defenders of the Standard Interpretation have responded. We'll see that in the final analysis, the need to treat computational states as representations is left in serious doubt.

2.1 THE CCTC AND THE STANDARD INTERPRETATION

The central doctrine of the CCTC is that cognition is computation, which is itself to be understood as a form of quasi-linguistic symbol-manipulation done in accordance with specifiable rules. John Haugeland captures the basic idea this way: "Reasoning (on the computational model) is the manipulation of meaningful symbols according to rational rules (in an integrated system)" (Haugeland 1985, p. 39). Over the course of the last forty years, there have been many different theories and hypotheses that could be described as part of the CCTC. Many of these differ dramatically in styles of processing, information storage strategies, types of basic algorithms and representational forms. Yet as a group they share the core idea that cognitive systems carry out various tasks by shuffling, connecting, storing, repositioning, comparing, and in other ways maneuvering primitive and complex symbol tokens.

By now, this way of describing the CCTC has become something of an extended slogan, and, like many slogans, it is far from clear just what it all is supposed to mean. For instance, what does it mean to say a computational system does "symbol manipulation"? Who or what manipulates the symbols and how does it know which manipulations to perform? What determines what the symbols are about, and how do they get their representational content? Moreover, how is such a process supposed to account for cognition?

Some of these questions have been largely ignored by advocates of the CCTC, or at least have been thought to express "mere implementation" matters. For instance, in most computers, symbol manipulations are carried out by some sort of central processing executive. This is typically a sophisticated device that not only keeps track of which operations need to be executed, but actually performs many of the operations as well. Yet the issue of how this is done in actual brains is generally passed over as something to be determined by neuroscience researchers. Other questions are seen to involve uninteresting technical details or even mere terminological quirks. For example, what it means to say that symbols are manipulated is just to say that symbols are erased, written and rewritten in various registers. This erasing and rewriting is what "symbol manipulation" refers to, along with the corresponding vernacular of symbol "shuffling," "combining," "rearranging," and such.[1] Thus, the CCTC claims that there is something like a neural code in the brain and symbols are "written" and "erased" by using this code.

In contrast to these somewhat neglected questions, other matters have been given far more attention and are properly treated as more central to the way the CCTC accounts for mentality. The most significant of these concerns the ways in which a symbol-manipulating paradigm is believed to provide compelling explanations for many of the central features of cognition and intelligence. Indeed, it is with regard to this matter that cognitive science has arguably enjoyed its greatest degree of success. Many would say that the CCTC has provided us, for the first time, with a scientifically robust account of human thought. Fodor describes the central "deep and beautiful idea" underlying the CCTC as "the most important idea about how the mind works that anybody has ever had" (Fodor 1992). Because there are now several superb introductions that describe the basic mechanics of the CCTC framework in considerable detail, what follows is a somewhat truncated synopsis of how the story generally goes.[2]

[1] See Copeland (1993), chapter 4, for an excellent discussion of how the erasing and writing actually get done in computers.

[2] See, for example, Block (1990); Boden (1977); Copeland (1993, 1996); Crane (1995); Harnish (2002); Haugeland (1985); Newell and Simon (1976); and Robinson (1992).

2.1.1 Levels of analysis and mechanized inference

The overarching explanatory picture of cognitive science is one that treats cognition as an input–output process that admits of different levels of analysis. This picture often reflects David Marr's conception of cognition as having three distinct levels of explanation or analysis (Marr 1982). At the top level is the specification of the explanandum – the cognitive task that we are attempting to explain. Marr calls this the "computational" level,[3] where the specification is typically an input–output function expressing the cognitive process or capacity, sometimes called the "task domain." These capacities are functions that can vary from converting numbers into products, sentences into parse trees, playing chess, constructing a medical diagnosis, and so on for a variety of things we think cognitive agents do. It is popular to regard the input as the proximal stimuli to a perceptual system and the output as some sort of motor movement, but the inputs and outputs can be any types of states that define a cognitive operation. In Marr's account, for example, the outputs were often also input to some other early stage of visual processing.

At the next lower level of analysis is what Marr referred to as the "algorithmic" level.[4] In truth, computational systems don't directly convert numbers or arrays of chess pieces. Instead, they convert representations of such things. On a standard computer, for instance, keyboard typings represent the input while the display on a computer screen (or perhaps some sort of print-out) represents the output (Cummins 1989). So at this level, we re-describe the explanandum in representational terms: The new explanandum is not the conversion of numbers into sums, but the conversion of representations of numbers (numerals) into representations of sums. In fact, this is the *general* form of cognitive science explananda, even for non-CCTC accounts like connectionism. What we want explained is how the brain performs this representational conversion. It is at the algorithmic level that theories like the CCTC provide us with a set of operations whereby symbols that designate the relevant inputs are transformed into symbols designating the appropriate output (where appropriateness is determined by the function specified at the top level). This is the level of analysis at which most traditional cognitive theories, like those of the CCTC, are pitched.

[3] This label is somewhat misleading since the sort of operations many regard as computations are not found at this level, but at the lower, algorithmic level.

[4] This same level is also sometimes called the "representational level," or, more controversially, the "syntactic level."

Finally, at the bottom level there is what Marr referred to as the "implementational" level. Here we find descriptions of the actual physical structures and processes that implement the operations occurring at the algorithmic level. It is at this level where physicalism is vindicated because psychological processes are mapped onto actual physical events and processes. In everyday computers, the explanatory posits of this level would be things like wires and circuits and their various causal interactions. In humans and other organisms, the relevant posits would be the actual neurological states, structures and processes that implement the cognitive operations that a given theory proposes.

As noted, in this three-tiered framework it is at the middle, algorithmic level where the CCTC theories attempt to explain the kinds of processes that account for mentality. It is at this level where the critical symbol-manipulations take place. Roughly, the accounts put forth are descriptions of processes in which symbols are shuffled about and combined so that their transformations produce appropriate input–output conversions. The processes are typically assumed to implement an algorithm, whereby complex operations are broken down into simpler and more basic operations – all performed on symbolic tokens. Thus, the explanatory strategy is task-decompositional; to understand a sophisticated cognitive process or capacity, the CCTC posits a series of sub-systems that perform different sub-tasks. We can continue breaking sub-systems down into simpler components until we reach the sort of processes that can be described by implementational elements. The process is familiar to those who build and program computers, but the goal in cognitive science is not to build a smart computer. Instead, the goal is to use our understanding of computers to model and help explain what is alleged to be taking place in our heads when we perform some cognitive chore. Advocates of the CCTC have called our brains "Physical Symbol Systems" (Newell and Simon 1976; Newell 1980) because of the crucial role symbols play in this way of understanding cognitive processes. Formal symbols are essential because they serve as the central elements (or "medium") of the type of algorithmic operations specified by the theory. Orderly symbol manipulations are physically possible because, as computers demonstrate, symbols can be erased, rewritten, stored, combined by purely mechanical processes. As long as the symbols are encoded so that their physical form (sometimes called their "syntactic" properties or their "shape") can be exploited by other elements of the system, these symbolic transformations can give rise to input–output mappings of the sort we are trying to explain. Put simply, the CCTC is the overarching explanatory framework that attempts to account for cognition by appealing to this kind of symbol-manipulating process.

The explanatory strength of this framework can be seen more clearly when we recognize that physical symbolic operations of this type can correspond to the states and transformation of formal systems. For instance, if we include some connectives and quantifiers (whereby symbols like ">" stand for the material conditional), then the CCTC approach tells us how a physical system can implement a deductive system. Upon receiving the symbolic input "P," a mechanical system can search and retrieve the data structure of the form, "P > Q," and then, on the basis of those tokens and the system's physical architecture, write and register the new token, "Q." In fact, it shows us how formal operations can be mechanized that not only adhere to the basic principles of deductive logic, but more complex formal systems like predicate calculus. It is this feature – their ability to mechanically implement logical processes – that gives actual computers their powerful problem-solving capacities. Computers are often called "number crunchers," but a better description would be "logic implementers." Applying this understanding of computation to psychology, we get at least part of a theory of how our mind/brain works. Cognition is seen as rule-governed symbolic modifications that are possible because the system is sensitive to the physical form of the symbols. Physical systems – including the brain – can thereby use physical symbols to become a sort of inference-machine.

2.1.2 *The CCTC, RTM, and the Standard Interpretation*

At the end of the last section, I described the purely mechanical symbolic operations put forth by the CCTC as "inferences." However, if we want to account for *actual* inferences of the sort we would recognize as thought processes, then the story, as it stands, won't do. Mechanical operations involving merely formal tokens won't suffice for an account of what we normally think of as *reasoning*. To be an account of reasoning (or inference, decision-making, thinking, etc.), the relevant units of computation must be representational states like beliefs and desires. What's missing from the account just presented are states with full-blown representational status. A critical question then, is how do we come to view formal computational symbols in this way?

The Standard Interpretation suggests an answer to this question. The critical move in constructing a folk psychological interpretation of the computational processes is to exploit the parallelism that is possible between the causal relations among computational symbols, on the one hand, and the causal and inferential relations assumed to exist between beliefs and other propositional attitudes on the other hand. As we saw in the last chapter, folk psychology treats propositional attitudes as states that play different causal roles. The Standard Interpretation expands on this idea and claims that causal roles distinguish not just propositional attitudes from other mental states, but also different sorts of propositional attitudes. A given representation playing one sort of role is a belief state; the same representation playing a different sort of role is a desire state. The idea that mental processes are mediated by mental representations has been called the "Representational Theory of Mind," or RTM for short. While there are different variants of RTM, most versions regard psychological processes as the interplay of representational states in accordance with formal principles. Some of these principles govern the way primitive representations can be conjoined to form complex representations, whereas others determine the causal relations between representational states. What matters here is that on the Standard Interpretation, RTM helps to wed commonsense psychology to the CCTC. If the inferential relations of the attitudes are casual relations that can be mirrored by the symbols posited by CCTC, then we can treat the symbols as beliefs and explain ordinary inferences and judgments. If we say the symbol "P" represents that it's currently raining while the symbol "Q" represents that I need to drive to work, and ">" represents the material conditional, then the mechanical operation described above – where a token of "P" and a token of "P > Q" led to a tokening of "Q" – can explain how a person decided to drive to work. It is by mirroring the inferential and causal relations of beliefs and the like that the CCTC can provide intuitively plausible accounts of everyday reasoning. If we can do this for a variety of different cognitive capacities and skills, then we have what amounts to a robust theory of ordinary thought.

The picture presented here is often described as one whereby a physical system implements a syntactic engine which drives a semantic engine. It is an appealing model of cognition because it preserves much of what we take for granted about mental states and processes while at the same time showing how these states and processes might be physically realized. Perhaps the best known advocate of this account is Jerry Fodor. Fodor argues that defending the commonsense assumptions about thoughts on the one hand, and, on the other hand, defending the central framework of RTM, and, by extension, the CCTC, all amounts to the same basic enterprise (1987, p. 16). According to Fodor, a theory of cognition provides an endorsement of commonsense psychology if it posits inner states that are semantically evaluable, have causal powers and abide by the generalizations of folk psychology. The inner symbols of the classical model are the perfect candidate for this explanatory role. As he puts it, "[c]omputers show us to connect semantical with causal properties for *symbols*. So, if having a propositional attitude involves tokening a symbol, then we can get some leverage on connecting semantical properties with causal ones for *thoughts*" (Fodor 1987, p. 18).

The Standard Interpretation, then, offers answers to two important, closely related questions about the nature of representation in the CCTC. The first question is: What type of representational notion is invoked in computational explanations of cognitive processes? The answer proposed is, the same type of representational notions that are invoked by commonsense psychology. Computational processes are seen as a mechanized version of folk psychological reasoning, and this is possible only if the symbols being manipulated are viewed as analogues of familiar commonsense mental representations. The symbols of the CCTC are realizers of beliefs, desires, concepts, and other mental representations put forth by our pre-theoretical framework. The second question is: What is it about the CCTC's explanations of cognition that motivates the positing of representations? And the answer, once again, stems from the Standard Interpretation's assumption that the explanation of cognition is in large measure an explanation of the states and processes put forth by folk psychology. Computational symbols come to be regarded as the representational states of folk psychology so that the explanations provided by the CCTC are pertinent and "make sense."[5] The symbols earn this treatment by virtue of their capacity to causally replicate the intuitive ways we assume thoughts interact to produce mental processes.

As many have noted (Fodor 1987; Pylyshyn 1984; Clark 2001), one of the virtues of this perspective is that it appears to provide a compelling vindication of our commonsense understanding of the mind. With the Standard Interpretation, the threat of eliminative materialism – the view that there are no such things as beliefs and other commonsense mental representations – can be significantly reduced, since the CCTC is seen to be fully committed to propositional attitudes. At the same time, the CCTC also inherits a considerable degree of intuitive plausibility from folk psychology. If we identify the elements of computational systems as the sorts of states we intuitively think are at the heart of cognitive processes, as the

[5] See Pylyshyn (1984).

Standard Interpretation does, then this intuitive understanding of the mind will make the computational perspective seem *prima facie* plausible.

To sum up, on the Standard Interpretation, the CCTC offers an account of cognition that makes use of certain notions of representation. The notions invoked are the same notions of mental representation put forth by commonsense psychology, especially our notion of belief. The symbols serve as representations in this theory by serving as the computational equivalent of these ordinary mental representations. The CCTC thereby provides us with a framework that vindicates folk psychology by providing us with symbolic posits that causally interact in the manner of thoughts and other mental representation. But at the same time, folk psychology helps vindicate the CCTC by providing intuitive support for some of the "folksy" aspects of symbolic representations. The end result is a reading of the CCTC whereby the notion of representation employed is, roughly, the same as our folk notions of mental representation.

2.2 DIFFICULTIES WITH THE STANDARD INTERPRETATION

Despite the overall attractiveness of this perspective, I think it is fundamentally wrong-headed. The Standard Interpretation proposes what I believe to be wrong answers to both of the questions posed earlier: It is not the case that the basic notions of representation employed in the CCTC are folk notions of mental representation, nor is it the case that the motivation for treating computational symbols as representations is so that commonsense psychology can be given a scientific home. In the next chapter, I'll try to provide what I think are the right answers to these questions. But for the remainder of this chapter, I want to explain how the Standard Interpretation leads to trouble. The problems stem from an aspect of the job description challenge presented in the last chapter; in particular, the need to explain how a given structure or state actually serves as a representation. Because the Standard Interpretation claims that computational symbols function as analogues for propositional attitudes, it ignores what is distinctive about CCTC explanations that makes the positing of representations a more integral aspect of the theory. While the symbols can behave in a manner that mirrors the intuitive syntactic and causal nature of mental representations, this is not enough to demonstrate that whatever content they have is explanatorily relevant.

While the specific version of the job description worry I have in mind hasn't received a great deal of attention, something close to it has loomed in the background of more traditional challenges to computational representationalism. It will help, then, to frame our discussion of these matters by looking at two of these challenges and especially the manner in which defenders of representationalism have responded. One challenge, presented by Searle, is that computational symbols lack the sort of intentional content ascribed to propositional attitudes. According to Searle, this entails that they lack the status of real representations. The second challenge, endorsed by Stich, contends that folk psychological notions of representation are not suited for serious scientific psychology. From this Stich argues that we should adopt a strictly syntactic interpretation of computational psychology and abandon the representational perspective altogether. Both of these challenges exploit the idea that computational representationalism is rooted in folk psychology, that computational symbols qualify as representations by serving as analogues to beliefs and other folk notions of mental representation. While proponents of the Standard Interpretation have vigorously responded to both Searle and Stich, these responses, we'll see, raise as many questions as they answer. In an oft-quoted passage, Haugeland suggests that a motto of the CCTC is, "if you take care of the syntax, *the semantics will take care of itself*" (Haugeland 1985, p. 106). But as these critics point out, it is far from clear how, exactly, the semantics is supposed to simply "take care of itself."

2.2.1 *Searle's criticism of computational representation*

The author who has done the most to argue that in computational systems the semantics will *not* "take care of itself" is Searle (1980). His Chinese Room argument has become such a prominent landmark on the cognitive science terrain that it hardly needs reiteration. A man is locked in a room with a box of Chinese symbols he doesn't understand, but is also given an instruction manual (written in a language that he does understand) that tells him how to manipulate the symbols in various ways. After some Chinese characters are passed along to him, he consults the manual and carries out a number of elaborate symbolic operations, focusing only on the shape of the symbols. Eventually, he hands back through an output slot a different set of Chinese symbols. Unbeknownst to him, he has just been asked some question in Chinese and has responded in a perfectly reasonable manner. How does he do it? By implementing an algorithm that is designed to create the impression that he (or the system) really understands Chinese. But, Searle argues, this is only an illusion; he has no idea what the symbols actually mean. In fact, from the standpoint of the system itself, the symbols have no real representational status – they mean nothing. According to Searle, merely implementing a formal program is not sufficient for real mentality or understanding; the syntactic operations of the computational model do not provide us with an adequate account of cognition.

As some have noted (Warfield 1998), there are really two distinct conclusions that Searle uses the Chinese Room example to establish. Both involve the claim that running a program is in some way inadequate, but they differ with regard to what, exactly, running a program is inadequate for. The first conclusion is that computational processes are inadequate for instantiating understanding or mentality in general. While the man in the room specifically lacks understanding of the Chinese language, Searle is clearly making a more general point that reaches beyond linguistic knowledge. His intuitive point, and the one that makes the argument so compelling, is that no amount of formal symbol manipulation alone can ever give rise to real comprehension or understanding of anything. The man could instantiate a symbol-manipulating program for any cognitive process and neither the man nor the running program nor the entire system would actually instantiate that cognitive process. This is Searle's rejection of Strong AI, the idea that running the right program is sufficient for having a mind. Searle's Chinese Room can be seen as a counter-example to the metaphysical claim that any system that runs the right program automatically becomes a thinking system.

Searle's other conclusion more directly challenges the representational character of the CCTC. Searle notes that computational systems succeed at their various tasks by manipulating symbols by virtue of the syntactic properties and not by virtue of any sort of semantic content. In his Chinese room, the man simply checks the shape of the characters, checks the instructions, and then shuffles the symbols about by virtue of these two factors. By contrast, *real* mental representations – things like our thoughts and ideas – intuitively interact with one another and produce behavior by virtue of what they are about. This leads Searle to deny that computational symbols are representational states at all, since they lack any real representational content. At most, they have the sort of derived intentionality we discussed in the last chapter – a sort of meaning that is assigned by us, the outside programmers and observers. Since the symbols lack the intrinsic intentionality of real thoughts, there is no sense in which they serve as representations *for* the system itself. In short, Searle argues that it is simply wrong to regard the symbolic structures that mediate computations as representations. As he puts it, "syntax alone is not sufficient for semantics, and digital computers insofar as they are computers have, by definition, a syntax alone" (1984, p. 34).

While both of Searle's conclusions are important,[6] it is the second thesis that directly challenges the Standard Interpretation of the CCTC. It does so by pulling the rug out from under the idea that classical symbols provide a suitable reduction base for the representational posits of folk psychology. Searle's argument is similar to our earlier point that a theory of cognition cannot simply adopt folk notions of representation because those notions involve features, like intrinsic intentionality, that need further explication. You don't get such an explication just by positing states that causally interact in the manner put forth by the CCTC. Searle's claim is not simply that the meaning of computational symbols does no explanatory work, or that the content of the symbols is causally inert. His point is that computational symbols don't have any real content, and consequently computational symbols aren't really representations at all. Searle's argument challenges the very idea that classical computational theory is itself a *representational* theory of cognition. He claims it isn't, because the symbols in classical architectures don't have the original intentionality associated with real thoughts.

Responding to the Chinese Room argument has become a minor industry in the philosophy of psychology, and in the next chapter I'll offer my own rebuttal. But for now the important point to note is how both Searle and the defenders of the CCTC argue from the perspective of the Standard Interpretation and link the explanatory value of computational representations with our commonsense understanding of the mind. Most supporters of the CCTC agree with Searle that the posits of the CCTC need to be sufficiently similar to folk mental representations if they are going to serve as representations. So the debate is not over the *sort of* representational posit put forth by the CCTC. Instead, it is over whether or not the CCTC succeeds in accounting for all of the important aspects of that posit – in particular, whether it succeeds in accounting for the intentional content that we associate with beliefs and other mental representations.

Many popular strategies for responding to Searle concede that, by itself, the CCTC does *not* account for the sort of intentionality needed to treat computational symbols as representations. There is, in other words, a bit of back-pedaling on the idea that "the semantics will take care of itself." If positing representations entails positing belief-like states, and if positing belief-like states entails explaining intentional content, then the CCTC, as such, doesn't actually posit inner representations! But according to

[6] On Searle's own account of intentionality, these conclusions are closely related since he holds that the content of thoughts is closely linked to consciousness and our background understanding.

defenders of the Standard Interpretation, this only means something more needs to be added to convert the inner symbols into real representations. On the Standard Interpretation, while there is nothing about the way the symbol shuffling succeeds in producing appropriate input–output conversions that suggests the symbols actually serve as representations, we nevertheless *can* view them as representations if we supplement the theory with a workable theory of content. For example, a popular response that Searle himself considers – the "Robot Reply" – claims that the CCTC needs to be supplemented with some sort of story that connects the inner symbols to the world via the right sort of causal links. With this proposal, if the inner computational symbols get hooked to the world in the right way, then they actually will have the sort of original intentionality ascribed to folk mental representations, and can thereby serve as robust representational states.

Before we evaluate this response to Searle's argument, it is worth pausing to consider what it suggests about the explanatory connection between the CCTC and representation. For starters, it would seem to clearly undermine the basic idea that the CCTC explains cognition by invoking inner representations. If it is generally agreed that the computational processes can be fully understood without an appeal to any sort of intentional content, then at most what the CCTC provides is the non-representational chassis or framework whose internal states can be *converted into* representations with some added story about how states get their meaning. That is, the CCTC merely provides the non-representational precursors to (or non-representational vehicles for) the representational states posited by folk psychology. As one author puts it, "... if you let the outside world have some impact on the room, meaning or 'semantics' might begin to get a foothold. But, of course, this response concedes that thinking cannot be simply symbol manipulation. Nothing can think simply by being a computer" (Crane 2003, p. 128). Moreover, if the explanatory role of classical symbols is to provide a vehicle for vindicating the representational posits of folk psychology, then this is certainly an odd sort of vindication. In the last chapter, it was noted that the representational posits of folk psychology are assumed to have their content essentially – beliefs and other propositional attitudes are individuated by virtue of their content. Insofar as computational symbols, *qua* theoretical posits of the CCTC, don't, as such, have any sort of content, then the CCTC fails to capture the most central aspect of what is supposedly being vindicated. Even if the symbols can *acquire* semantic properties through causal links to the world, the need for this is independent of the explanatory framework provided by the CCTC. It's not that the representational character of the CCTC's theoretical posits establishes a scientific home for beliefs; rather it seems that, at least in part, the desire to find a scientific home for beliefs drives the representational characterization of CCTC theoretical posits.

These considerations allow us to see how the Standard Interpretation of the CCTC leads down a path that gives rise to skepticism about representation in the CCTC. The Standard Interpretation assumes that representations are needed in the CCTC in order to account for the sort of thought processes recognized by folk psychology. It thereby emphasizes the degree to which computational symbols can, by virtue of their syntactic properties, mirror the causal relations of the attitudes. Hence, there is a parallelism between the causal activity of computational data structures and the causal activity commonsense assigns to folk mental representations. But the parallelism alone is not enough to bestow representational status on the symbols; an additional theory of content is needed. Consequently, the symbols' status *as* representations is not, on the Standard Interpretation, something built into the explanatory framework of the CCTC. If positing representations entails positing belief-like states, and if positing belief-like states entails explaining intentional content, then the CCTC, as such, doesn't actually posit inner representations![7]

Let's return to the question of whether or not the standard-plus response to Searle is ultimately successful. Recall that the strategy is to argue that if we can construct an adequate theory of content for computational symbols (one that conforms with our assumptions about ordinary thought content), then we would have all the necessary ingredients for a robust and complete representational theory. Searle denies that such a theory of content has been given, arguing that (for example) the sort of head-world causal links thought to account for content fail to actually do so. But let's set aside the debate over content and focus instead on whether the proposed strategy, if successful, would provide what is needed. On the Standard Interpretation, the CCTC allows us to meet the job description challenge by revealing how something can *function as* mental representations. They function as mental representations by functioning in the way commonsense psychology assumes beliefs and other propositional

[7] Dan Weiskopf has pointed out that if we adopt a functional role semantics, it might be said that the causal roles associated with computational symbols also provide the content for the symbols, so a content story is built into the functional story of the CCTC after all. Yet putting aside the many difficulties associated with functional role semantics, most functional role semanticists concede that syntactic relations cannot be the sole determinant of content for mental representations (Block 1986). Moreover, such an account of content would still leave unexplained what it is about the symbols that makes their role a representational one.

attitudes function in the mind. So if what we want is an account that accommodates our ordinary notion of mental representation, what we wind up with is an account with two parts. One part, provided by the CCTC, is an account of how inner states acquire the causal/functional properties associated with specific sorts of propositional attitudes. The other part, some sort of theory of content, is an account of how the same states come to have the original intentionality associated with mental representations. Put the two parts together and we have, it would seem, an account of the mind that is fully representational and vindicates folk psychology. If having the belief that the Dodgers won is a matter of the mental representation with the content "the Dodgers won" playing a certain functional role (i.e., the believing role), then even though the CCTC does not itself provide a theory of content, it does provide an account of what it would be to serve as a mental representation of this sort. Or does it?

The problem with this picture is that it doesn't allow us to explain what it is for something to serve as a representational state, *simpliciter*. On the Standard Interpretation, the causal-syntactic framework of CCTC leaves that part of the job description unexplained, since it presupposes that data structures playing these different roles are already functioning as a mental representation. What the CCTC shows is how a state serving to represent that the Dodgers won comes to function as a belief or desire. It enables us to explain when the structure is functioning as a belief-type representation and distinguish this from when it is functioning as a desire-type representation. What it doesn't show, on the Standard Interpretation, is how (explanatorily prior to all this) the state in question comes to serve as the representation the Dodgers won. This latter role, whatever it is, would be one that representations serving as different propositional attitudes all *share*. On the Standard Interpretation, the CCTC provides a causal story that allows us to distinguish a believing representation from a desiring representation, etc. It doesn't give us an account that allows us to distinguish inner representations, as such, *from everything else*.

Recall that the job description challenge requires that when a representational posit is invoked in a theory, there is some sort of account of how the state is supposed to function as a representation in the system in question. The problem with the account we've been considering is that, even after the account is supplemented with a theory of content, we still don't have a sense of how inner states are supposed to serve as representations in computational processes. We do get an account of the sorts of causal relations that are involved in making something a representation of a certain sort (analogous to the relations that allow us to distinguish a hand-held compass from one that is mounted on an automobile's dash). This is provided by the causal/syntactic architecture of CCTC that explains the difference between beliefs, desires and other propositional attitudes. And we can assume we get some sort of account of content-grounding conditions (analogous to the nomic dependency between the needle's position and magnetic north). This might be provided by a theory linking the inner symbols to their intentional objects through some sort of causal relation. But what we don't get is an account of the computational (or physical) conditions that explain why a state or structure should be treated as serving as a representation in the first place (analogous to someone using the compass needle's position to discern the location of magnetic north). What is missing from the Standard Interpretation is a story about how symbolic posits actually serve as representational states in the sort of functional architecture proposed by the CCTC. It isn't provided by the syntactic or causal interactions of the symbols because these causal relations are of the wrong sort. They account for the differences between types of inner representations, but they don't account for the way in which a computational system uses data structures *as* representations. And it isn't provided by proposed content-grounding conditions, such as causal links between the representation and its intentional object, because those conditions generally don't, by themselves, *make* something into a representational state.[8] It is often assumed that the combination of the inner causal relations with the content-bestowing head-world links provides all that is needed. But we can now see that we need something more. The missing feature isn't constituted by these other factors, just as a compass's role as a representational device isn't constituted by its being mounted on the dash and its causal connection to magnetic north. What's missing is an account of exactly what it is about the way the brain uses inner symbols that justifies regarding those inner states as representations.

Representations are thus similar to other special science posits that involve multiple dimensions. Consider what it is for some piece of metal to serve as a form of currency. To fully understand a quarter's role as a bit of currency, it would not be enough to describe some of its unique causal relations, such as what it does in a Coke machine. Nor would it be sufficient to explain the process that bestows upon the quarter its value of 25 cents. To be sure, these matters are relevant to our understanding of the coin's status as currency. But one could understand various causal

[8] A similar point is made by Adams and Aizawa (1994) in their criticism of Fodor's account of representation.

interactions and even the process whereby its value is stipulated, and still be ignorant of *how* it actually serves as a unit of exchange (say, by not knowing that it is physically traded for goods). Representations are similar. Besides standing in some sort of content grounding relation and besides participating in a number of other causal relations, they are, more fundamentally, employed to stand for something else. How this happens is left unexplained by the Standard Interpretation.

The problem, as I see it, stems from the way the Standard Interpretation ties representation in the CCTC to commonsense psychology. Because the motivation for treating symbols as representations is connected to the explanatory value of folk psychology, it is seen to stem from considerations that are distinct from the explanatory scheme used by the CCTC. Consequently, there is a blind spot to the way in which that explanatory scheme actually *does* explain how computational symbols serve as representations. The debate prompted by Searle's Chinese Room argument helps reveal that blind spot and show how, on the Standard Interpretation, representation is presented as something that really *isn't* an inherent part of CCTC explanations. In the next chapter, I'll argue that, in fact, when we adopt the correct interpretation of the explanatory role of representations in classical models, we actually *do* get an account that makes the notion of representation a central and indispensable explanatory posit. On the proper treatment of CCTC, the notion of representation invoked can be seen to meet the job description challenge. But before we see how that story goes, it will pay to look at another challenge to computational representationalism.

2.2.2 *Stich's criticism of computational representation*

Stich's challenge to representational computational psychology is somewhat more nuanced than Searle's, but it is perhaps more damaging to the Standard Interpretation. In his 1983 book "From Folk Psychology to Cognitive Science," Stich adopts the Standard Interpretation tenet that treating computational explanations as involving representations amounts to treating computational explanations as committed to the posits of folk psychology. While he thinks this may be possible (unlike Searle), he also thinks it is a bad idea. Folk psychology individuates propositional attitudes by virtue of their content. However, according to Stich, there are a number of reasons for thinking that content-based taxonomies are ill-suited for serious, scientific psychology. Because content-based taxonomies are based upon head-world relations, they will individuate mental states in ways that cross-classify more scientifically respectable taxonomies, such as those based upon causal powers or causal roles of the internal state. Moreover, Stich maintains that content ascriptions for propositional attitudes are based upon similarity judgments – we ascribe beliefs by imagining what we would say in situations similar to the subject's. Thus, content ascriptions involve a high degree of vagueness, are highly parochial and context sensitive, and fail with subjects who are too dissimilar from ourselves, such as the very young and the mentally ill. In effect, Stich denies the alleged parallelism between the causal/syntactic operations of computational symbols on the one hand, and the content-based descriptions of mental processes presented by folk psychology on the other. Because representational psychology is, for the most part, identified with folk psychology, and because folk psychology makes for lousy science, Stich argues that we need to drop the appeal to representations in CCTC altogether.

In place of a representational cognitive science, Stich argues psychologists should employ a purely causal/physical, or syntactic theory – one that individuates mental states by appeal to their purely non-semantic properties. In other words, Stich agrees with Searle that computational symbols should not be treated as surrogates for folk psychological posits like beliefs. But unlike Searle, Stich does not argue that the CCTC should be abandoned. Instead, it should be re-conceived so that internal symbolic structures are treated as formal tokens individuated by virtue of their purely syntactic properties. Stich's so-called "Syntactic Theory of the Mind" is what he claims computational psychology should look like – a theory that retains the same basic architecture of classical computational theory, but makes no attempt to treat the inner states as representations. The purely syntactic generalizations and taxonomies provided by such an outlook are, according to Stich, much better suited for a scientific psychology because they carve the inner computational machinery at its causally salient joints. So Stich can be seen as making a negative point and a semi-positive point. The negative point is that since a representational perspective relies on a folk psychological framework, and given that such a framework leads to taxonomies that are ill-suited for science, computational psychology should reject representationalism. The semi-positive point is that it is okay for the CCTC to abandon representationalism since a non-representational, syntactic account works just fine in accommodating CCTC explanations.[9]

[9] Though he can hardly be characterized as an anti-representationalist, Fodor has pressed similar points in his well-known paper, "Methodological Solipsism" (Fodor 1980). Fodor endorses what he calls the "formality condition" which states that computational theories should individuate symbolic states without appeal to semantic properties. Here, Fodor appears to join Stich in insisting that treating CCTC symbols as representations is, at best, gratuitous.

Although Searle's argument is clearly very different from Stich's, it is worth pausing to consider the ways in which their views overlap. Both Searle and Stich agree that classical computational accounts of cognition should not be treated as representational. Why not? Because, according to these authors, computational symbols fail to serve as an adequate reduction base for folk notions of representation like belief. Searle concludes from this, "so much the worse for classical computationalism." Stich concludes from this, "so much the worse for folk psychology." Both authors assume (at least tacitly) that the question of whether or not the classical model of cognition provides us with a workable notion of representation is to be answered by focusing on our notions of belief, desire and other propositional attitudes. Searle argues that computational states would need to be more belief-like to qualify as real representations; since they aren't, computationalism is a flawed model of the mind. Stich appears to agree that the representational status of computational states rests on their being belief-like, and since he thinks belief-like states are not scientifically respectable, representationalism is rejected. Despite their criticism of the marriage between computational psychology and folk psychology, both authors share the Standard Interpretation's assumption that positing classical computational representations amounts to positing belief-like states in a computational framework. In fact, they use this assumption to challenge the link between CCTC and RTM.

While defenders of the Standard Interpretation respond to Searle by arguing that it is *possible* to convert symbolic structures into belief-like representations, the challenge posed by Stich forces them to explain why it is *necessary*, or at least beneficial, to do so. This is important for our discussion because it goes right to the heart of the issue of the explanatory value of representations. If Stich's challenge can be answered and it can be shown that our understanding of cognitive processes is significantly enhanced by treating CCTC symbols as propositional attitudes, then the Standard Interpretation would itself be vindicated. We would then have reason to suppose that the folk psychological notion of representation in the CCTC has some degree of explanatory legitimacy. On the other hand, if it should prove quite difficult to answer Stich's challenge and if the benefit of treating CCTC symbols as beliefs and desires is left in doubt, then that can be taken to indicate one of two possibilities. It could be taken to show, as Stich suggests, that CCTC shouldn't be regarded as a representational theory. Or, it could be taken to show that the Standard Interpretation is wrong to equate CCTC representations with folk psychological states, and that a different way of thinking of computational representations is in order.

Because, as we've noted, most authors today endorse the idea that the content of mental representations can be reduced to some other set of naturalistic conditions, many would reject Stich's analysis of content ascription, especially his suggestion that they are based merely on similarity judgments. Most accounts of content determination hold that the semantic content of our thoughts is determined by objective factors (like certain sorts of causal relations) and that a scientific psychology could adopt content-based taxonomies that are neither parochial nor context-sensitive. However, even with these more robust accounts of content ascription, many writers would concede that the syntactic approach offers a more causally accurate picture of the inner workings of the mind. Given that the syntactic properties are what actually determine the causal role of symbols in computational processes (the ones that contribute to the causal powers of the symbols), a syntactic taxonomy would be more descriptive and would have greater predictive power. So why not regard the symbols as simply non-representational tokens that serve as the medium of purely mechanical computational operations? This is the key question Stich poses for CCTC representationalists.

There have been two popular strategies for responding to this challenge and defending representationalism. One strategy is to directly assail Stich's argument for the Syntactic Theory by showing there is a flawed assumption in the argument. The second strategy is to appeal to some explanatory desideratum that is achieved through representationalism and would be lost if we were to adopt a purely syntactic framework. Each can be seen as a way to support the union between the CCTC and folk psychology for those who adhere to the Standard Interpretation.

A common version of the first strategy offers a sort of self-refutation argument against Stich's assumption that taxonomies more finely tuned to the details of the inner workings of some system are superior to those that are more coarse-grained and abstract. The problem with this assumption, goes the response, is that it applies to the syntactic level itself. After all, there are levels of description of brain processing that are purely physicochemical, or perhaps even molecular, that would provide a more detailed analysis and offer more accurate generalizations than the syntactic level. Thus, if Stich's reasoning is sound, we should abandon the syntactic level (and computational psychology altogether) and adopt the physicochemical level of description. But this seems much too drastic.[10] Block calls this argument the "Reductionist Cruncher" and states, "... the physicochemical account will

[10] See, Bickle (2003) for an account that does not treat this as absurd.

be more fine-grained than the syntactic account, just as the syntactic account is more fine-grained that the content account ... if we could refute the content approach by showing that the syntactic approach is more general and more fine-grained than the content approach, then we could also refute the syntactic approach by exhibiting the same deficiency in it relative to a still deeper theory" (Block 1990, p. 280).

While Block's argument offers a valuable lesson for many sorts of reductionism, its application to Stich's challenge is misplaced. Block seems to assume that Stich is advocating a switch from a higher level of analysis to a lower one, where levels of analysis correspond to different levels of organization or composition in physical reality. On this view, the kinds and entities described at a lower level constitute the kinds and entities described at a higher level. Block's complaint is against those who would try to abandon descriptions of higher levels of composition in favor of theories and descriptions at a lower level (abandoning biology, say, for physics). But Stich is concerned with something very different; namely, with determining which properties of the symbolic entities at a single level of organization (the one appropriate for computational psychology) we ought to use when constructing our psychological taxonomies. The transition from a content-based taxonomy to a purely syntactic taxonomy does not involve a transition in levels of analysis or organization. Stich isn't arguing that we should abandon the computational level of analysis at all. Rather, he is challenging the idea that at the algorithmic level of analysis, the posits should be classified by appealing to their alleged semantic properties as opposed to their syntactic properties. The issue is one concerning *types* of taxonomies, not *levels* of taxonomy.

What is really in dispute concerns which properties matter for understanding the sort of job the posits of CCTC actually perform. The debate between Stich and representationalists is about the proper job description for computational symbols. Stich argues they should be viewed as formal tokens that mediate syntactic operations and processes. Block and others argue they should instead be viewed as representations. But because the Standard Interpretation leaves it unclear exactly how data structures are serving as representations in computational processes, it is unclear what it is about computational explanations that warrants a representational job description for the symbols. Some have argued[11] that a syntactic taxonomy will miss generalizations captured by a content-based approach. But it is important to bear in mind that taxonomies and generalizations are cheap. It is, after all,

[11] See, for example, Pylyshyn (1984).

easy to construct various taxonomies for symbolic data structures that are more abstract than those based on semantic properties. If we can vindicate the representational treatment of computational symbols simply by showing that it allows unique generalizations, then we can vindicate *any* taxonomy of computational states since they all allow for unique generalizations. What needs to be shown, then, is not that a content-based approach to computational psychology invokes vocabulary, captures generalizations, and makes classifications, etc. that are *different* from what would be provided by a purely syntactic, non-representational approach. Of course it does. Instead, what needs to be shown is that the vocabulary, generalizations and classifications allowed by the representational approach buys us something worth having. We need some justification for thinking that the generalizations provided by the representational story provide something of significant value that would be missing from a purely syntactic story.

This brings us to the second strategy for responding to Stich's challenge, which is to appeal to some further explanatory desideratum that, allegedly, only the representational picture can provide. When asked for this, the most common response is to invoke some sort of principle of *rationality*. Here it is claimed that the representational picture provides (and the syntactic account leaves out) a framework that allows us to characterize cognitive processes as rational. Perhaps the most explicit expression of this is found in Pylyshyn (1984), so it is worth quoting him at length:

> What I am claiming is that the principle of rationality . . . is a major reason for our belief that a purely functional account will fail to capture certain generalizations, hence, that a distinct new level is required (p. 34) . . . [I]n a cognitive theory, the reason we need to postulate representational content for functional states is to explain the existence of certain distinctions, constraints and regularities in the behavior of at least human cognitive systems, which, in turn, appear to be expressible only in terms of the semantic content of the functional states of these systems. Chief among the constraints is some principle of rationality. (1984, p. 38)

Arguments similar to this have been presented by others.[12] The basic sentiment can be summarized as follows:

(1) Theories of cognition must account for the rationality of cognitive systems.
(2) The only way we can account for the rationality of cognitive systems is by treating internal states as representations.
(3) Therefore, computational theories of cognition must treat inner symbols as representations (and not merely syntactic or functional states).

[12] See, for example, Fodor (1987) and Rudder-Baker (1987).

So here, it seems, we have a clear answer to Stich's challenge. We need to view cognitive systems as rational, and to view them as rational, we need to treat their inner states as representations, which means, on the Standard Interpretation, as commonsense mental representations. Moreover, this could perhaps be used to handle our primary job description concern of explaining how computational symbols actually *serve* as representations. They serve as representations by virtue of serving as not just symbolic tokens, but as states in a system engaged in rational inferences. To explain cognition, we need to account for rationality, and to account for rationality, the symbols posited by the CCTC must function as commonsense mental representations.

Unfortunately for defenders of the Standard Interpretation, this line of reasoning has a number of shortcomings. It is far from obvious exactly what Pylyshyn means by "rational," and there are certainly notions of rationality that could be explained without any appeal to inner representations. For example, if all that is meant by "rational" is that the behavior is appropriate given the circumstances, or that the inner processes provide the right input–output mappings, then it is hard to see why a syntactic model couldn't provide this. In fact, Stich himself offers a point-by-point comparison between, on the one hand, a content-based account of someone being told her building is on fire and subsequently fleeing, and, on the other hand, a purely syntactic story of the same cognitive process and behavior (Stich 1983).[13] Since the syntactic story explains the same life-preserving behavior as the representational account, it seems the syntactic theory can explain rational behavior as well.

Of course, many would argue that to be seen as rational, the behavior itself must be given an intentional description (answering a phone call, fleeing from danger, etc.), and that *this* requires that the behavior arise from inner states characterized in intentional terms. While this is a huge topic that demands of book-length treatment, let me briefly offer some reasons why I reject this perspective. First, many of the arguments for this view rest on dubious claims about analytic definitions for terms used to characterize the relevant behavior. For example, it might be claimed that

[13] Here's a sample of how Stich develops his account: "Why did Mary come running from the building? It's a long story. First, she had a long standing D-state whose syntactic form was that of a conditional, viz. F ⊃ L (corresponding to the desire to leave the building if it is on fire), where F and L are themselves syntactically well-formed strings ... Mary began to inhale smoke ... The direct consequence of inhaling smoke was that Mary came to have a B-state (corresponding to the belief that she was inhaling smoke). From this B-state and the long-standing B-state I ⊃ N (corresponding to the belief that if one is inhaling smoke then there is a fire nearby) she inferred (i.e., was caused to add to her B-store) a token of N ..." (1983, pp. 174–175).

for us to characterize an action as "fleeing from danger," it is necessary for us to assume that the subject believe that there is danger nearby, has a desire to avoid danger, knows how to flee, etc. But it seems clear that our basic understanding of what it is to flee from danger could survive, perhaps with modification, if we dropped representational descriptions of inner causes. In fact, Stich's own syntactic analysis of the woman escaping the burning building shows us how this might go. Second, as Fodor (1987) has noted, there are deep problems associated with the idea that we should individuate behaviors in ways that are sensitive to the content of the underlying psychological states. Insofar as content depends upon relational factors that most would deny are determinants of behavior (such as linguistic affiliation), it is better to classify behaviors in ways that pertain to the causal powers of inner states. That amounts to adopting the syntactic theory when describing the inner causes of behavior. Third, it seems just false that an ascription of rationality requires an intentional characterization of behavior. "Crouching low," for example, is arguably a non-intentional description of behavior, but surely this can be seen as fully rational in certain situations. Finally, we've seen that what cognitive theories typically try to explain is not simply overt behavior, but cognitive capacities characterized as the transformation of input representations into output representations. There is no reason provided by the Standard Interpretation to think the intermediary states need to be representations for these input–output mappings to successfully instantiate some sort of function (like addition). But if these input–output mappings correspond with some cognitive capacity in a systematic way, then it certainly seems they would qualify as rational.

But what about the computational processes themselves? If we want to regard those inner processes as rational inferences, don't we need to treat them as involving inner representational states? Once again, the matter depends on what we mean by "rational." If we take it to simply mean that the processes are in accordance with the rules of deductive and inductive logic, then the answer is clearly "no." We can instead treat the symbols as formal tokens whose causal relations follow the basic rules of logical transformations. After all, logical rules just are rules that govern formal interactions between uninterpreted variables. If the syntactic operations mirror valid logical schema, then it would certainly seem that this notion of rationality doesn't require representational states with content.

However, on another interpretation of "rational" – the one Pylyshyn and others presumably have in mind – a purely formal, syntactic interpretation of computational processes won't do. The reason is that to qualify as rational in this second sense, the relations between inner elements need to adhere to a different set of rules than those of formal logic. They would need to adhere to the "laws" and generalizations that folk psychology assigns to propositional attitudes and other folk mental states. This is the notion of rationality associated with explanations like, "She left the building because she thought it was on fire" or "He went to the fridge because he wanted a beer and thought there were more left." In other words, to be rational in this sense is to be driven by inner processes that involve states generally recognized as *reasons*. What sort of states are these? Commonsense mental states like beliefs, desires, hopes, fears, and other propositional attitudes, which are, of course, representational states. Consequently, for the computational process to be rational in this sense, we must regard the participating states as mental representations. The argument presented earlier needs to be modified to look more like this:

(1) Theories of cognition must account for the rationality of cognitive systems.
(2) To be rational (in this sense) is to instantiate the processes described by commonsense psychology.
(3) To instantiate the processes described by commonsense psychology is to instantiate processes involving commonsense mental representations like beliefs and desires.
(4) Therefore, theories of cognition must appeal to commonsense mental representations like beliefs and desires.
(5) Therefore, computational theories of cognition must treat inner symbols as commonsense mental representations (and not merely syntactic or functional states).

Initially, this might seem to bring us closer to what the Standard Interpretation needs – an argument that establishes what is gained by treating computational symbols not just as representations, but as representations of the sort recognized by commonsense psychology. CCTC is committed to representational states because it is committed to explaining rationality, and rationality in this context is just defined as processes involving belief, desires and so on. Yet a bit of reflection reveals that this won't work as a response to the sort of challenge Stich is offering. To respond to Stich's syntactic theory, we would need to show what is wrong with a non-representational interpretation of the CCTC, and the suggestion is that it will miss certain "distinctions" and "regularities" associated with a principle of rationality. But when we ask what this means, it turns out that these are just the distinctions and regularities that come along with treating states as commonsense mental representations. In other words, the reason we need to treat computational states as propositional attitudes is so that we can treat computational processes as the sort of processes that involve propositional attitudes. This is hardly a convincing argument against a skeptical challenge to representationalism. We noted above that the sort of distinctions and generalizations captured by a representational perspective will be different from those offered by a syntactic account. Yet what needs to be shown is not that the representational framework is distinct, but that it is distinct in a way that is substantially superior (in terms of explanatory and predictive power) from the syntactic framework. The suggestion is that representationalism is indeed superior because it allows us to view computational processes as rational. This is what is supposed to be gained by treating the CCTC symbols as belief-type states. But being rational on the current proposal just means treating CCTC symbols as belief-type states! Thus, the argument is circular.

So the defender of the Standard Interpretation who appeals to rationality winds up with the following dilemma. Either we can define what it is for a system to be rational in a way that makes it distinctive, say by appealing to some formal system like deductive logic. But if we go this route, the relevant states can be treated as merely non-representational tokens and the syntactic account works just fine. Or, we can define rationality in a way that makes representations essential – indeed, makes beliefs and desires essential. But then being rational amounts to the same thing as implementing a system with the sort of representational states invoked by folk psychology. If you are a serious skeptic about the value of folk psychology, or of representation in general, rationality in this sense becomes something scientific psychology shouldn't care about. Either way, a principle of rationality doesn't help those who claim that data structures in CCTC systems need to be treated as representations.

In fairness to Pylyshyn and others, it is far from obvious that the appeal to rationality is based upon the same understanding of the explananda of cognitive theories that we have been assuming throughout this discussion. Here, I've characterized a central explanandum of cognitive theories as various cognitive capacities and skills defined in terms of representational input–output functions. We want to know how minds convert representations of chess boards into representations of moves, or how representations of sentences are converted into grammaticality judgments. Under this assumption, it is an open question whether the cognitive machinery responsible for these conversions involves folk psychological states like beliefs and desires – or, for that matter, inner representational states of any sort (apart from the inputs and outputs). It is at least possible, under this construal of cognition, that a theory of the mind that denies inner representations could prove to be true. However, there are other conceptions of the explananda of cognitive science, and one of them takes it as a given that cognitive processes are of the sort (more-or-less) suggested by folk psychology. On this construal of cognition, one of the things we want explained are ordinary inferences like those mentioned above – e.g., someone deciding to leave a building because she believes it is on fire and wants to avoid injury. Folk psychological processes are, from this perspective, part of what we want theories like the CCTC to explain, and this would include an account of the rationality (and, presumably, irrationality) of these thought processes. To accomplish all this, computational symbols must be treated as realizers of propositional attitudes.

While I think there are a number of problems with the idea that cognitive scientists should assume (or do assume) that folk psychology accurately captures the nature of the mind, and that they should (or do) thereby treat commonsense psychological processes as their explanatory target, I'm willing to grant that sometimes this is the case. Even if this is so, this assumption brings us right back to our earlier point that the CCTC, as understood by the Standard Interpretation, fails to provide the proper reduction base for commonsense notions of mental representations. The reason is that on the Standard Interpretation, there is no account of how computational symbols actually serve as representations in computational processes. While there may be a theory of some sort of content-grounding relation for the symbols (like head-world causal relations) and also an account of the inner causal relations that explain how symbols can play the role of different propositional attitudes (distinguishing the believing role from the desiring role, etc.), we don't get an account of what it is for the symbols to actually function as representations in the type of operations presented by the CCTC.

Since I am ultimately going to argue that there actually *is* an explanatory pay-off in treating CCTC symbols as representations, I won't pursue this matter further here. The critical point is that, once again, the problem is due to the Standard Interpretation of the CCTC. I've suggested that the Standard Interpretation comes with the tacit assumption that we can show how symbols are representations by claiming that they realize or instantiate propositional attitudes. But as should now be clear, this doesn't work. You can't use the fact that A is the proposed reduction base for B to establish that A has all the relevant features of B. That is, you can't make computational symbols function as representational states by proposing that they serve as the things with which folk mental representations are identified. Rather, one needs to first establish that computational symbols serve as representations in computational explanations of cognition, irrespective of their role as possible reducing posits of folk representations. Then, once we've established that symbols do indeed play such a role, we can ask whether or not they might be the sort of thing that instantiates beliefs and desires. In the next chapter, I'll argue that, on the proper interpretation of the CCTC, this latter strategy is indeed possible. Once we recognize that the notions of representation at work in CCTC are essential to the type of explanatory framework the CCTC provides, we don't need any *further*, folk psychological justification for treating them as representations. But to see things that way, we need to abandon the Standard Interpretation.

2.3 SUMMARY

In this chapter I've tried to do two things. First, I've presented what I take to be a very popular way of thinking about representation in CCTC that pervades the literature and dominates many discussions and debates. That way – what I've been calling the Standard Interpretation – suggests that CCTC is, by and large, a mechanized version of mental processes as conceived by commonsense psychology. Hence, the notion of representation ascribed to CCTC is seen as the same notion put forth by folk psychology. Second, I've tried to show how this outlook leads to problems for representationalism in classical cognitive science. These include the concession that content (and hence, representation) is not actually an element of computational explanations of cognition, along with a failure to explain what it is for computational symbols to serve as representations (even if an account of content is added). A common slogan is that classical computation requires a medium of representations. But on the Standard Interpretation, this looks like a bluff – there appears to be little reason to think a medium of *representations* is needed, as opposed to a medium of syntactically distinct tokens.

In the next chapter, I'll argue that all of this is due to a faulty understanding of CCTC and the notion of representation it employs. The Standard Interpretation is not the proper way to look at these matters, and when we gain a better understanding of the way computation is thought to explain cognition, we can also see why and how representation is needed. The notions that are needed, however, have little to do, at least directly, with the notions of mental representation found in folk psychology. Instead, they are theoretical posits that are as important and central to the CCTC as are notions like *algorithm* or *sequential processing*. Hence, I'll argue that CCTC employs notions of representation that allow for a clear explication of what it is to serve as a representation, and one that shows exactly what would be missing from Searle's Chinese Room and Stich's syntactic theory of the mind. We've seen how *not* to understand representation in classical cognitive science; now let's see how it ought to be understood.

3

Two notions of representation in the classical computational framework

In the last chapter, we saw how representation in the CCTC is commonly regarded. In this framework, representation is generally treated as closely linked to our commonsense conception of the mind and, in particular, to our understanding of propositional attitudes. We also saw how this perspective fails to provide an adequate account of why the CCTC needs to appeal to representations at all. If the Standard Interpretation was the only interpretation, we would have little reason to suppose that there is any real explanatory pay-off in treating the posits of classical AI as standing for something else.

But the Standard Interpretation is not the only way to look at things. In this chapter I want to present another perspective on the CCTC, one that I think reveals why the classical framework provides a legitimate home for a robust notion of internal representation. Actually, my claim will be that there are two related notions playing somewhat different but nonetheless valuable explanatory roles. One notion pertains to the inputs and outputs of computational processes which help to define the cognitive task being performed. As we'll see, given the sort of explanatory strategy usually adopted by the CCTC, this also provides a notion of *inner* representation as well. The second notion pertains to data structures that in classical explanations serve as elements of a model or simulation. That is, according to many theories associated with the CCTC, the brain solves various cognitive problems by constructing a model of some target domain and, in so doing, employs symbols that serve to represent aspects of that domain. Since other authors have already provided detailed explications of these representational notions, my goal will be to provide an overview and, where necessary, perhaps modify or extend these earlier analyses.

Both of the notions of representation I am going to defend in this chapter have been criticized as suffering from serious flaws. One alleged problem, related to concerns discussed in the last chapter, challenges the idea that these representational notions are sufficiently robust to qualify as *real* representations, as opposed to merely instrumental or heuristic posits. A second worry is that the account of content connected to these notions is unacceptably indeterminate between different possible interpretations. I plan to demonstrate that once we appreciate the sort of explanatory work these notions are doing, we can see that their alleged shortcomings are actually much less serious than is generally assumed. Both notions are quite robust, and while there is indeed an issue of indeterminacy associated with them, it doesn't have any bearing on the explanatory work they do in the CCTC. I should say up front, however, that I have fairly modest goals in this chapter. I do not intend to address all of the various problems and challenges that have been raised (or could be raised) in connection with these notions of representation (in fact, I doubt if such an exhaustive defense is possible for *any* representational posit). My aim is simply to show that there are notions of representation in the CCTC that are not based on folk psychology, that are essential to the explanatory strategies offered by the CCTC, and that can handle some of the more basic worries associated with naturalistic accounts of representation. If I can demonstrate that the CCTC posits internal representations for good explanatory reasons, then I will have accomplished my primary objective.

To show all this, the chapter will have the following organization. First, I'll provide a sketch of each notion of representation and show how it does valuable explanatory work in the CCTC. Then, I'll consider two popular criticisms against these notions – that they are merely useful fictions and that the associated theory of content is plagued with rampant indeterminacy. I'll argue that both criticisms can be handled by paying close attention to the way these notions are actually invoked in accounts of cognition. Finally, there are a number of side issues that it will help to address for a more complete picture. In the final section, I offer a brief discussion of each of these important side issues.

3.1 IO-REPRESENTATION

In the last chapter, we saw how Marr's model of cognitive science involved three levels of description and how the "top" level involved the specification of a function that more or less defines the sort of cognitive capacity we want explained. Consider again a simple operation like multiplication. Although we say various mechanical devices do multiplication, the transformation of numbers into products is something that, strictly speaking, no physical system could ever do. Numbers and products are abstract entities, and physical systems can't perform operations on abstract entities. So, at the algorithmic level we posit symbolic representations of numbers as inputs to the system and symbolic representations of products as outputs. We re-define the task of multiplication as the task of transforming numerals of one sort (those standing for multiplicands) into numerals of another sort (those standing for products). The job of a cognitive theory is to explain (at this level of analysis) how this sort of transformation is done in the brain.

In fact, this general arrangement, whereby the explanandum is characterized as the conversion of representational inputs into representational outputs, will apply to most approaches to cognitive explanation. This is simply because cognitive processes themselves are typically characterized as an input–output conversion couched in representational terms. Pick any cognitive capacity that you think a scientific psychology should attempt to explain, and then consider how it should be characterized. For example, take the ability to recognize faces. The input to any cognitive system that recognizes faces will not be actual faces, of course, but some sort of visual or perhaps tactile representation presented by the sensory system. The output will also be a representation – perhaps something like the recognition, "That's so-and-so," or perhaps a representation of the person's name. Or consider linguistic processing. The challenge for most cognitive theories is *not* to explain how an event characterized in physiological terms (say, eardrum motion) brings about some other event characterized in physiological terms, but rather, how an acoustic input that represents a certain public-language sentence winds up generating a representation of, say, a parse-tree for that sentence. A theory about how the visual system extracts shape from shading is actually a theory about how we convert representations of shading into representations of shape. The same general point holds for most of the explananda of cognitive science. Indeed, this is one of the legitimate senses in which cognitive systems can be viewed as doing something called "information processing." While automobile engines transform fuel and oxygen into a spinning drive-shaft, and coffee-makers convert ground coffee to liquid coffee, cognitive systems transform representational states into different representational states.

Given the sort of analysis I am offering, an immediate question that arises about these types of input–output representations concerns the way they meet the job description challenge. In what sense do they function *as* representations, not just for our explanatory purposes, but for the actual cognitive system in question? There are two possible answers that could be offered. The first is to avoid the question altogether and say that the question is outside of the domain of cognitive theorizing. Cognitive theories are in the business of explaining the processes and operations that convert input representations into output representations; the concern of these theories (and therefore my analysis) is with the nature of *internal* representations. The nature of the input and output representations that define cognitive operations (and thereby define psychological explananda), while perhaps an important topic, is not an important topic that is the primary concern of cognitive modelers. Theoretical work has to start somewhere, and in cognitive science it starts with an explanandum defined in this way.

However, while there is some truth to this answer, it is as unsatisfying as it is evasive. A second and better (though admittedly controversial) answer is to say that there is considerable evidence that minds do certain things, and one of the main things they do is perform cognitive tasks properly described as the transformation of types of representations. It appears to be a fact of nature that certain minds can do multiplication, recognize faces, categorize objects, and so on. Well, what does that mean, exactly? It means that the cognitive system in question can convert, say, representations of numbers into representations of their product, or perceptual representations of an object into a verbal classification. The states that are the end-points of these processes are thereby serving as input–output representations for the cognitive system in question. The end-points serve as representations not because cognitive researchers choose to define them that way, but because we've discovered that cognitive systems employ them that way, given the sorts of tasks they actually perform. Below, I'll return to this question as it pertains to an *internal* sort of input–output representation. For now, the key point is that we are justified in treating a cognitive system's inputs and outputs as representations because, given what we know about cognitive systems, we are justified in characterizing many of their operations as having certain types of starts and finishes; namely, starts and finishes that stand for other things.

Cummins offers this explanation of the input–output notion:

> For a system to be an adder is for its input–output behavior to be described by the plus function, +(<*m*, *n*> = s. But + is a function whose arguments and values are numbers, and whatever numbers are, they are not states or processes or events in any physical system. How, then, can a physical system be described by +? How can a physical system traffic in numbers, and hence add? The answer, of course, is that numerals – that is, representations of numbers – can be states of a physical system, even if the numbers themselves cannot . . . The input to a typical adding machine is a sequence of button pressings: <*C*, *A1* + *A2*, = >, that is, < clear, first addend, plus, second addend, equals >. The output is a display state, *D*, which is a numeral representing the sum of the two addends. (Cummins 1991, p. 92)

Cummins calls this the "Tower-Bridge" picture of representation, because it involves two levels of transformations – physical and, in the case of

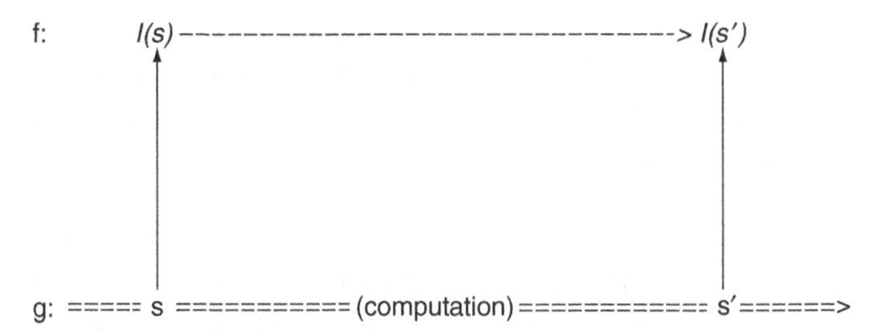

Figure 3a Cummins's proposed Tower-Bridge picture of computation (1991). The top level is the cognitive task being explained (f), the bottom level (g) is the algorithmic level of computational processes. The vertical arrows correspond with the interpretation of the bottom level input and output symbols, s and s′. Reprinted by permission from MIT Press.

addition, mathematical, which are conjoined on either end by semantic links between the physical representations and the things they stand for. Schematically, the picture is illustrated in figure 3a.

In much of his writing, Cummins characterizes this notion of representation as *the* notion employed in the CCTC. Because connectionist accounts also appeal to representations as the inputs and outputs of their networks, this leads him to the surprising conclusion that the CCTC and connectionists use the same notion of representation. This outlook is correct if we only consider the way both theoretical frameworks adopt similar specifications of psychological explananda. However, it is important not to confuse theory-neutral specifications of the explananda with the internal explanatory posits of particular cognitive theories. Since cognitive processes are defined with representational states as their end-points, it is a mistake to treat this notion of representation as *belonging to* the CCTC, or invoked *by* the CCTC. Since most theories treat types of input–output transformations as their starting point, the input and output themselves are not part of any particular theory's explanatory apparatus.

Nevertheless, a very similar sort of representational notion *does* play a critical role in the CCTC. This becomes clear once we look *inside* of cognitive systems as they are understood by the CCTC accounts. As we saw in the last chapter, sophisticated cognitive capacities are typically explained by the CCTC by supposing that the system is composed of an organized system of less sophisticated sub-systems. By decomposing complex systems into smaller and smaller sub-systems, we can adopt a divide-and-conquer style of explanation whereby the performance of complex tasks is explained by the performance of increasingly simpler tasks (Fodor 1968; Cummins 1975, 1983; Dennett 1978). As Cummins puts it, "psychological phenomena are typically not explained by subsuming them under causal law, but by treating them as manifestations that are explained by analysis" (1983, p. 1). Task-decompositional explanations are the norm in the CCTC, and they give rise to the popular "flow-chart" style of explanatory theory. It is this conception of cognitive systems that requires us to posit representations that serve as the inputs and outputs for the inner sub-systems that comprise the CCTC account. Internal mini-computations demand their *own* inputs and outputs, and these representations that are external to the mini-computation are, of course, *internal* to the overall system.

Task-decompositional analysis is a popular explanatory strategy in several different domains (like biology), yet theories in these domains don't all appeal to internal representations. So why are internal representations necessary for functional analysis when we are dealing with cognitive systems? The answer stems from the way the sub-systems and sub-routines in computational processes are typically understood. A general assumption of the CCTC is that many of the tasks performed by the inner sub-systems should be seen as natural "parts" of the main computations that form the overall explanandum. That is, they should be defined as procedures or sub-routines that are natural steps in a process that instantiates the more sophisticated capacity that is ultimately being explained. Our ability to do multiplication, for example, might be explained by appealing to a sub-process that repeatedly adds a number to itself (Block 1990). But to view the sub-process in this way – as a sort of internal mini-computation – then we need to regard *its* inputs and outputs as representations as well. If there is an inner sub-system that is an adder, then its inputs must be representations of numbers and its outputs representations of sums. If these internal structures are not serving as representations in this way, then the sort of task-decompositional analysis provided by the CCTC doesn't work. We won't be able to view the sub-system as an adder, and hence we won't be able to see how and why its implementation is essential to the overall capacity being explained. Consequently, certain structures that are internal to the system – structures that serve as inputs and outputs of certain intermediary sub-systems – must be seen as functioning as representations of matters that are germane to the overarching explanandum.

This point has been made with different terminology in Haugeland's classic treatise on cognitivism (1978). Haugeland introduces the notion of an intentional black box (IBB), which is (roughly) a system that regularly produces reasonable outputs when given certain inputs under a systematic interpretation of the inputs and outputs. Haugeland suggests that an information processing system (IPS) should be viewed as a type of intentional black box that lends itself to a further analysis. Such an analysis usually involves an appeal to IBBs that are internal to the IPS – i.e., a task-decomposition of the larger system into smaller sub-systems. A crucial feature of this type of explanation, then, is that certain internal states are interpreted as representing facets of the task in question:

> Moreover, all the interpretations of the component IBBs must be, in a sense, the same as that of the overall IBB (=IPS). The sense is that they all must pertain to the same subject matter or problem . . . Assuming that the chess playing IBB is an IPS, we would expect its component IBBs to generate possible moves, evaluate board positions, decide which lines of play to investigate further, or some such . . . Thus, chess player inputs and outputs include little more than announcements of actual moves, but the components might be engaged in setting goals, weighing options, deciding which pieces are especially valuable, and so on. (1978, p. 219)

To avoid confusion, I'll refer to input–output representations that make up the explanandum of the cognitive theory as "exterior" inputs and outputs, and input–output representations that help comprise the explanans of the CCTC as "interior" inputs and outputs. Interior input–output representations are a sub-system's own inputs and outputs that are internal to the larger super-system's explanatory framework. Since it is not uncommon to have nested computational processes, the sub-system itself may have *its* own internal representations, which are themselves the inputs and outputs of a sub-sub-system operating inside the subsystem in question. Hence, being "exterior" and "interior" is always relative to the system under consideratation.

We can now see that Cummins's Tower-Bridge picture needs augmentation. In between the two main end-point spans, there should be several internal bridges with end-points defined by their own mini-towers, linking internal physical states (the interior input–output representations) to aspects of the target domain that they represent. A more accurate portrayal of the CCTC would be something like what is presented in figure 3b,

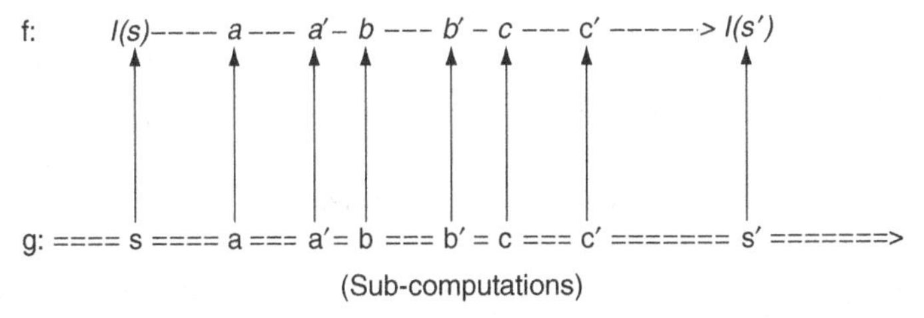

Figure 3b Cummins's Tower-Bridge diagram modified to accommodate inner computational sub-routines and representational states.

where the letters a, b, c correspond to the symbolic inputs of internal processes, while a′, b′, and c′ correspond to the representational outputs.

For our purposes, the most important aspect of this notion of representation is how it succeeds in meeting the job description challenge in a way that was not met on the Standard Interpretation. At least initially, we can see how interior input–output notion (or, the "IO notion") reveals how symbols *serve as* representations, given the hypothesized organizational architecture of the system. Data structures serve as representations because that is how the internal sub-systems treat them, given *their* job descriptions (e.g., performing addition, assessing chess moves, etc.). Serving as a representation of some feature of a target domain here amounts to serving as the sort of input or output required by a sub-processor solving a problem related to that domain. The content of the representation is critical for this role because unless the symbol stands for the relevant computational argument or value, it is impossible to make sense of the sub-system as a computational sub-system doing *its* job. Because it is an important element of this style of explanation, the interior IO notion of representation is not directly based on our folk notions of mental representation. We may come to view these inputs and outputs as thoughts, but the motivation to treat them as internal representations is not dependent upon our doing so. Even if folk psychology had never posited mental representations, the CCTC would still need to invoke interior IO-representations, given its explanatory framework. Yet while IO-representations don't accord with our commonsense understanding of *mental* representations, they nevertheless play a functional role that is intuitively representational in nature. It is an intuitively representational role because we recognize that systems doing things like addition, or comparing chess moves, treat their inputs and outputs as symbols standing for things like numbers or chess game scenarios.

Thus, the CCTC invokes a notion of internal representation that, contrary to what is implied by the Standard Interpretation, is actually built into the fabric of its explanatory framework and thereby does essential explanatory work. We can see this better if we briefly reconsider the criticisms of representationalism offered by Searle and Stich. The IO notion doesn't answer all of Searle's concerns about content and computational symbols. But consider the claim that there is *no* sense in which the symbols in the Chinese Room serve as representations for the system. On Searle's own account of the Chinese Room, the room does manage to provide appropriate answers to sophisticated questions about various topics. Suppose, in keeping with our algebraic theme, the questions asked are about the product of various numbers. So the input to the

Chinese room would be questions like "What is 3 x 7?," only written in Chinese. How does the room always manage to produce the right answer? According to the CCTC and Searle, the system does this by symbol manipulations that instantiate some sort of program. Let's assume the program is one that involves a sub-routine whereby one of the multiplicands is added to itself repeatedly.[1] We cannot understand this explanation unless we recognize that the man in the room's manipulations are, unbeknown to him, an adding process. And we cannot understand these manipulations as an adding process unless we recognize that Chinese characters generated by this process are serving as representations of sums. Putting it another way, we can't even make sense of how the symbol manipulations in the Chinese room succeed in generating the appropriate responses without invoking interior IO-representations. It doesn't matter that the person or thing manipulating the symbols doesn't understand what it is doing, or that the symbols lack the sort of intentionality associated with our thoughts. What matters is that we have an explanatory strategy that breaks a complex task (in this case, multiplication) into smaller tasks (i.e., addition) whereby the smaller tasks, by their very nature, require their inputs and outputs to be representations.

A similar point applies to Stich's anti-representationalism. Since on the Standard Interpretation, representational content appears to be superfluous to the CCTC type of explanations, Stich argues that the CCTC could get along just fine without it. But Stich's analysis is built on the assumption that the notions of representation at work in computational explanations are those derived from folk psychology. It neglects the possibility that there are notions of representation built into the sort of explanatory scheme adopted by the CCTC that need to be invoked for such a scheme to work. If we were to adopt the Syntactic Theory, avoiding all talk of representation and content, we would also be forced to abandon the type of task-decompositional explanation that is central to classical cognitive science. Since we couldn't treat the symbols as interior IO-representations, we couldn't

[1] The details might work as follows. After checking to see if one of the input characters represents either "0" or a "1," in which case special instructions would be followed, the man in the room is instructed to pick one of the input symbols and find its match on both the vertical column and horizontal row of what is actually an addition table. The syntactic equivalent of the other symbol is placed in a box. Once the symbol at the cross-section of the table is found (which would be the sum of one of the multiplicands added to itself), yet another symbol, designated by the instructions, is placed in another box. This is the system's counter. The symbol at the cross section of the addition table is then used to match a further symbol on the horizontal column, and the process repeats itself until the symbols in the two boxes match. At that point, a symbol matching the intersection symbol is handed through the output slot.

understand how the system succeeds by breaking a large computational operation down into related sub-operations. We could, of course, employ a syntactic type of task-decompositional explanation. We could track the causal roles of the syntactically individuated symbols, and thereby divide the internal processes into syntactic sub-processes. But we wouldn't be able to make sense of these operations as computationally pertinent stages of the larger task being explained. It is both explanatorily useful and informative to see a sub-system of a multiplier as an adder. It is not so useful or informative to see it as a mere syntactic shape transformer.

In accounting for the IO notion of representation, I've leaned very heavily upon the sort of explanatory strategy employed in the CCTC. I've suggested that because the CCTC uses a task-decompositional strategy that treats inner sub-systems as performing computations, then we need to regard the inputs and outputs of those sub-systems as representations. But this raises an important question – does the task-decompositional strategy provide a reason to think the inputs and outputs *actually are* representations, or does it instead merely provide us with a reason to *want* or *need* the inputs and outputs of these internal processes to be representations. Does it, from a metaphysical perspective, show us what serving as a representation amounts to? Or does it rather, from an epistemological perspective, create a need to have things serving as representations be the inputs and outputs for the inner computations?[2]

This is a difficult question and, quite frankly, I have changed my mind about its answer more than once. My current view is that the CCTC is committed to a sort of realism about inner computational processes, and this in turn reveals how the IO-representations actually function as representations, independent of our explanatory concerns. To adopt the language of Millikan (1984, 1993), the sub-systems act as representation "consumers" and "producers." But it is actually more complicated than this. They are consumers and producers of representations in a way that helps make the symbolic structures consumed and produced into representations (just as our consumption of a substance is what makes it have the status of food). The admittedly rough idea, briefly discussed above, is that computational processes treat input and output symbolic structures a certain way, and that treatment amounts to a kind of job assignment –

[2] As Dan Weiskopf has put it, "we seem forced to suppose that IO-representations are indeed representations because their being so is constitutive of the thing being explained (a kind of cognitive processor, i.e., a representation transformer). This doesn't directly answer the job description question, since we still don't know what properties metaphysically constitute IO-representations being representations" (personal communication).

the job of standing for something else. While an adder is something that transforms representations of addends into representations of sums, there is also a sense, given this arrangement, in which representations of addends are those symbolic structures that an adder takes as inputs, and representations of sums are structures an adder produces as outputs. There exists, then, a sort of mutual support between computational processes and representational states of this sort, with neither being explanatorily prior. Serving as a representation in this sense is thus to function as a state or structure that is used by an inner module as a content-bearing symbol. The inner modules are themselves like the inner homunculi discussed in chapter 1, whose treatment of their input and output can be seen as a type of interpretation. If the brain really does employ inner mini-computers, then their operations and transformations are, to some degree, what makes their input and output symbols into something functioning in a recognizably representational fashion. Below, in section 3.3.2, I'll address further the question of whether or not we can say the brain actually is performing inner computations in an objective, observer-independent sense.

These are just some of the issues that a sophisticated account of IO-representation would need to cover, and a complete account would need to explain considerably more. Yet remember that my primary objective here is fairly modest. Rather than provide a detailed and robust defense of this notion of representation in the CCTC, I merely want to reveal how the kind of explanations offered by the CCTC makes the positing of internal representations an essential aspect of their theoretical machinery. My aim is to demonstrate that there is a technical notion of representation at work within the CCTC and to show how that notion has a considerable degree of intuitive and explanatory legitimacy. Although the notion has little to do (at least directly) with our commonsense understanding of mental representations, it has a lot to do with the kind of explanations provided by classical computation theories. Yet it is not the only notion of representation in the CCTC that answers the job description challenge, has intuitive plausibility and does important explanatory work. The other notion is related, but nonetheless involves a different sort of representational posit that does different explanatory work. We turn to that notion now.

3.2 S-REPRESENTATION

In the first chapter we discussed Peirce's three different types of signs, noting that one of these, his notion of icons, is based on some sort of similarity or isomorphism between the representation and what it represents. The idea that the representation relation can be based on some sort of resemblance is, of course, much older than Peirce and is probably one of the oldest representational notions discussed by philosophers. But there is also the related though different idea that there can be a type of representation based not on the structural similarity between a representation and its object, but between the system in which the representation is embedded and the conditions or state of affairs surrounding what is represented. A map illustrates this type of representation. The individual features on a map stand for parts of the landscape not by resembling the things they stand for, but rather by participating in a model that has a broader structural symmetry with the environment the map describes. A map serves as a useful and informative guide because its lines and shapes are organized in a manner that mirrors the relevant paths and entities in the actual environment. Given this structural isomorphism between the map and the environment, the map can answer a large number of questions about the environment without the latter being directly investigated. Of course, this is possible only if the specific elements of the map are treated as standing for actual things in the environment. The map is useful as a map only when its diagrams and shapes are employed to represent the actual things, properties and relations of some specified location. The same basic notion of representation is at work when we use models, such as a model airplane in a wind tunnel, or computer simulations of various phenomena. It is also at work when numerical systems are used to model real-world parameters or when geometrical figures are used to understand aspects of physical systems. These and other predictive/explanatory arrangements share with maps the core idea that some sort of structural or organizational isomorphism between two systems can give rise to a type of representational relation, whereby one system can be exploited to draw conclusions about the other system.

Along with Pierce, many philosophers have offered accounts of representation based upon these themes. For example, it forms an important part of Leibniz's theory of representation, where he tells us that representations involve "some similarity, such as that between a large and a small circle or between a geographic region and a map of the region, or require some connection such as that between a circle and the ellipse which represents it optically, since any point whatever on the ellipse corresponds to some point on the circle according to a definite law" (Leibniz 1956, pp. 207–208). More recently, Chris Swoyer has developed a more detailed general account of this type of representation, which he refers to as "structural representation" (1991). Swoyer makes an impressive stab at constructing a detailed formal analysis of this notion, but even more beneficial is his analysis of the kind of explanatory framework it yields, which he calls "surrogative reasoning." As Swoyer notes, when maps, models and simulations are used, we typically find out something directly about the nature of the representational system, and then, exploiting the known structural symmetry, make the appropriate inferences about the target domain. As he puts it,

> [T]he *pattern* of relations among the constituents of the represented phenomenon is mirrored by the pattern of relations among the constituents of the representation itself. And because the arrangement of things in the representation are like shadows cast by the things they portray, we can encode information about the original situation as information about the representation. Much of this information is preserved in inferences about the constituents of the representation, so it can be transformed back into information about the original situation. And this justifies surrogative reasoning . . . (1991)[3]

What does this have to do with cognitive science and the CCTC? While this notion of representation may not capture all of the ways in which computational processes are regarded as representations, it serves as an important, distinct, and explanatorily valuable posit of classical computational accounts of cognition. Just a quick survey of many well-known computational theories of cognition finds this representational notion repeatedly invoked in one form or another. This includes such diverse cognitive theories as Newell's production-based SOAR architecture (1990), Winograd's SHRDLU model (1972), Anderson's various ACT theories (1983), Collins and Quillian's semantic networks (1972), Gallistel's computational accounts of insect cognition (1998),[4] and many other types of CCTC accounts. Stephen Palmer (1978) presents an excellent overview of the many ways in which this type of isomorphism-based representation appears in classical cognitive theories. While Palmer notes that the form these representations take in different theories can vary widely, they all share a basic nature whereby "there exists a correspondence (mapping) from objects in the represented world to objects in the representing world such that as least some of relations in the

[3] Here Swoyer refers to the entire pattern as the structural representation, but in other spots he seems to treat the *constituents* of the patterns as representations. I'm inclined to adopt the latter perspective, though as far as I can tell, very little rides on this besides terminology.

[4] Gallistel tells us, "[a] *mental representation* is a functioning isomorphism between a set of processes in the brain and a behaviorally important aspect of the world" (1998, p. 13).

represented world are structurally preserved in the representing world" (Palmer 1978, pp. 266–267).[5]

Perhaps the main proponent of the view that cognition is computational modeling is Philip Johnson-Laird (1983). Echoing one of our general concerns, Johnson-Laird laments the fact that most symbol-based approaches to explaining cognition "ignore a crucial issue: what it is that makes a mental entity a representation *of* something" (1983, p. x). To correct this oversight, he suggests we need to understand the way cognitive systems employ mental models, and how elements of such models thereby function as representations. For Johnson-Laird, the idea that problem-solving is modeling applies even for what seem to be purely formal, rule-driven cognitive tasks such as deductive inference. He offers a compelling and detailed theory of different mental capacities that is built upon the core idea that computational states serve as representations by serving as elements of different models.

Besides Swoyer, the philosopher who has done the most to explain this notion of representation – especially as it applies to the CCTC – is Cummins (1989, 1991). Cummins calls this notion of representation "simulation representation." Since Cummins's simulation representation is sufficiently similar to Swoyer's structural representation, I'll stick with the conveniently ambiguous term "S-representation" to designate the relevant category. Cummins first explicates S-representation by noting how, following Galileo, we can use geometric diagrams to represent not just spatial configurations, but other magnitudes such as velocity and acceleration. While there need be no *superficial* visual resemblance between representation and what is represented (velocity doesn't look like anything), there is a significant type of isomorphism that exists between the spatial properties of certain geometric diagrams and the physical properties of moving bodies that allows us to use diagrams to make inferences about the nature of motion. It is this same notion that Cummins argues is at the heart of the CCTC. In other words, when classical computational processes are introduced to explain psychological capacities, this often includes an invoking of symbols to serve as S-representations. The mind/brain is claimed to be using a computational model or simulation, and the model/

[5] I am claiming that the classical computational framework has been the main home for a model-based conception of representation. In the next two chapters I'll argue that non-classical frameworks, like connectionism, employ different notions of representation, notions that fail to meet the job description challenge. But there are also a few connectionist-style theories that invoke model-based representations in their explanations. See Grush (1997, 2004) and Ryder (2004) for nice illustrations of such theories.

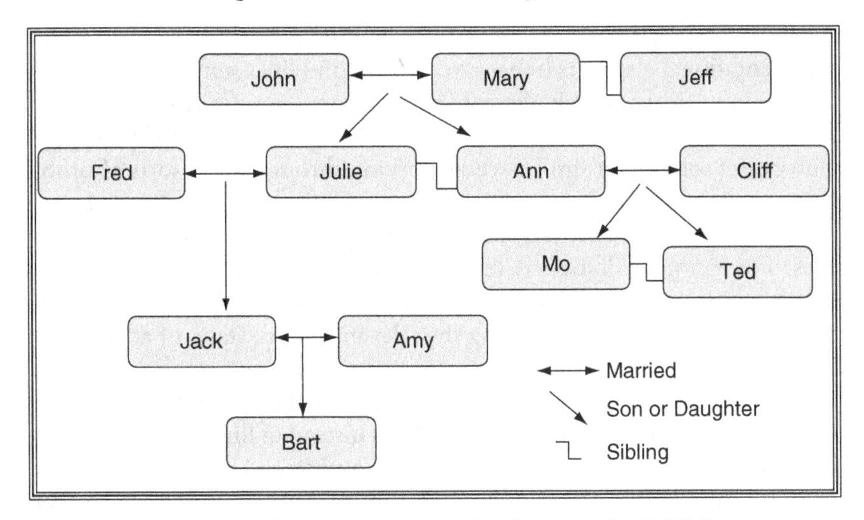

Figure 3c The family tree model used to determine familial links.

simulation is constructed out of symbols that are thereby serving as S-representations.

To get a better handle on how S-representation works in a CCTC system, it will help to step back and consider just how *we* might invoke similar sorts of representation to solve a problem. Suppose a person, Bob, is trying to determine whether two people are related, and if so, how. Bob knows many of the familial relations, but since the family is large he sometimes has trouble remembering how two people are related. So he gets a pen and a pad of paper and begins writing down the familial network. He does this by writing the name of each person, and then adding links to the names of other people, with each connecting link designating a specific type of relation (e.g., sibling, daughter/son, etc.). The result looks like the diagram in figure 3c. At times, Bob fills in blanks in his knowledge by making inferences about relations based on what he has already written (for example, he might come to realize that two people must be related in a way that had never before occurred to him). If two people are related in a certain way, then so and so follows, but if they are related in a different way, then something else follows. Eventually, Bob completes the diagram and then uses it to retrace the pertinent links and thereby establish how different people are related.

The manner by which Bob solves his problem is easy to see. He succeeds by constructing a *model* of real-world people and familiar relations which is then used to discover new facts. The relevant familial link between two people is discovered by exploiting analogous links in the model. The representational elements (the written names and lines standing for people and their relations) of his diagram re-create the specific real-world conditions he is seeking to learn more about. Moreover, we can easily imagine Bob doing something similar when working through other sorts of problems, including those where the pertinent relations are not familial, but causal, spatial, mathematical, modal or any of a variety of other possibilities. For example, if Bob is trying to work out what repercussions he should expect in light of certain events, he can once again use a pen and paper and draw a diagram linking the relevant events, states of affairs, and possible consequences. This time, instead of representing familial relations, the lines and arrows may represent causal or entailment connections between different propositions. Or perhaps instead of linking the pertinent elements with lines drawn with labels, he simply uses "if-then" statements to represent the relevant entailment relations. He might use a sketch that winds up looking more like a lengthy logical argument than a schematic, pictorial diagram. But it will arguably still be a representational model that invokes elements that serve to mirror the conditions and states of affairs and entailment relations that Bob is trying to understand. There will still be a type of isomorphism between the sketch and the target that can be exploited to learn certain facts about the target.[6] And in such an arrangement, elements of the model perform a certain job – they serve as representations of particular elements of the target domain that is being modeled.

Returning to cognitive science, the basic point that is generally ignored by the Standard Interpretation is that the CCTC is, by and large, a framework committed to the claim that when the brain performs cognitive operations, it does the same sort of thing as Bob. Of course, the CCTC doesn't claim the brain uses pen and paper. Instead, it uses the neural equivalent of a buffer or short-term memory device and some sort of process for encoding neural symbols. But just like Bob's diagram, the symbol manipulations alleged to occur in the brain allow for problem solving because they generate a symbolic model of a target domain. That is, the symbol manipulations should be seen as the implementing of a model

[6] Of course, questions about more abstract mappings and increasingly obscure forms of isomorphism loom large. We can imagine gradually transforming a map so that it no longer resembles any sort of map at all, and yet it still somehow encodes all of the same information about the relevant terrain. I'm willing to be fairly unrestrictive about what qualifies as a map, model or simulation, as long as there is a clear explanatory benefit in claiming the system *uses* the structure in question as such. See also Palmer (1978) and Cummins (1989).

or simulation[7] which is then used to perform some cognitive task. The symbols themselves serve as S-representations by serving as parts of the model. As Cummins puts it, "Representation, in this context, is simply a convenient way of talking about an aspect of more or less successful simulation" (1989, p. 95).

For example, many production-based systems, like Newell's all-purpose SOAR architecture (1990), function by invoking a "problem-space" of a given domain and then executing various symbolic operations or "productions" that simulate actual real-world procedures, thereby moving the system from a representation of a starting point to a representation of some goal state. If the system is trying to re-arrange a set of blocks (imagine it controls a robot arm), then it executes a number of operations on computational symbols that represent the blocks and their positions. By manipulating these representations in a systematic way, determined by the SOAR's own procedural rules, the system succeeds in constructing a model of the world that it can then transform in various ways that mimic real-world block transformations. To make sense of all this, we cannot avoid treating the various data structures of the computational architecture as representations of elements of the relevant problem-space. As Newell puts it, "This problem space is useful for solving problems in the blocks world precisely because there is a representation law that relates what the operators do to the data structure that is the current state and what real moves do to real blocks in the external world" (1990, p. 162). This sort of "problem-solving-by-model/simulation" is at the heart of the CCTC style of explanation. These processes are a mechanized version of what Swoyer referred to as "surrogative reasoning."

An obvious complaint about the analogy between Bob's use of the diagram and what goes on in classical computational systems is that Bob mindfully interprets the marks of his diagram (thereby bestowing them with meaning) while, as Searle would argue, the computational system has no idea what its symbols means. Isn't it right to say that Bob is using a representational system in a way that Newell's computational device *isn't*, given that Bob – but not the computer – is assigning meaning to the symbols?

[7] There may be significant differences between a model and a simulation, but here I will use these two terms interchangeably. In other words, I won't assume that there is a significant difference between a computer model of some phenomenon like a hurricane, and a computer simulation of the phenomenon. Some might say that models are static representations whereas simulations involve a *process*, but it seems there are plenty of uses of "model" whereby it designates a process as well; indeed, a computer model is just such a case.

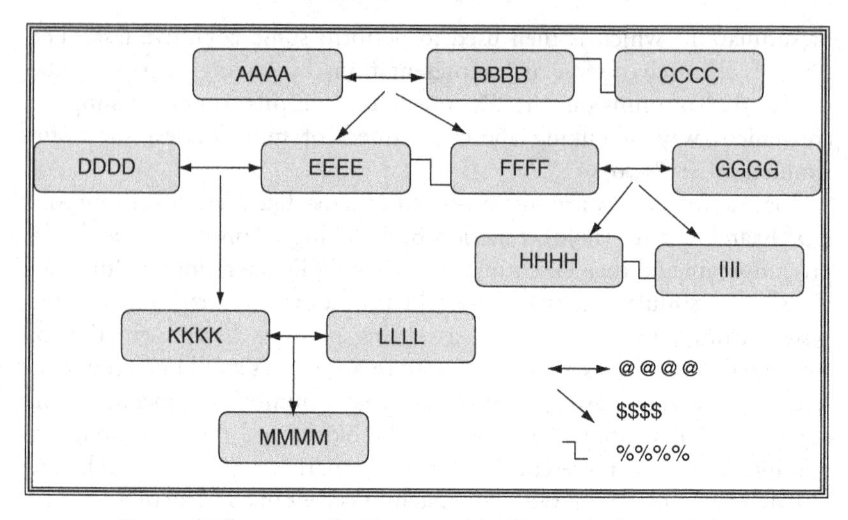

Figure 3d The opaque family tree model with meaningless symbols.

To answer this worry and get a better sense of the explanatory role of S-representation in the CCTC, we can consider what would happen to our explanation of Bob's problem-solving strategy if we were to substantially dumb him down and remove his own interpretive capacities. Suppose Bob doesn't understand what he is doing; the symbols are written in some language he doesn't comprehend, and he is simply following procedural rules that tell him such things as when to copy the symbols (by virtue of their shape) and when to look for matches. The diagram he is working with might now look like the diagram in figure 3d. A procedural rule might say, "If symbols AAAA and BBBB are connected by a line with arrows on each end, then put an X here." Bob has no idea that the letters stand for people, or that lines indicate different types of relationships. In this way, Bob becomes just like the man in Searle's Chinese Room – something that mindlessly manipulates syntactic structures in a manner that nonetheless generates solutions to some problem domain. Bob's use of the diagram becomes more like what might take place in a real computational device. The critical question now is this: Does making Bob more like an automated system substantially alter the representational role of the figures in his diagram? Or, alternatively, can we now explain and fully understand Bob's performance without ever characterizing the marks on the paper in representational terms?

On one reading of the question, what is being asked is whether or not Bob's sequence of operations is directly guided by the meaning or semantic character of the figures on his paper. With regard to this question, the answer is famously "no." As we saw with the Chinese Room, the features of the symbols that allow mindless-Bob (or computational central processing units) to recognize, organize, arrange, etc. the symbolic structures are the non-semantic "syntactic" features – the symbols' shapes. It is by virtue of the shapes of the symbols (and the traced lines between those symbols) that mindless-Bob is guided though his various operations, and to understand the purely mechanical nature of those operations we needn't treat the symbols as representing anything.

At the same time, however, if told only that familial relations were discovered through focused attention to shapes and marks on paper, we would find this explanation of Bob's performance grossly inadequate. As we saw with the IO notion, we would still want to know how he was able to achieve success. We would want to be told what this arrangement of marks could possibly have to do with a familial connection between two people, and how it is that making marks on a piece of paper, and focusing on their shape, could lead to a discovery of that relationship. If told nothing more than mindless-Bob drew a diagram according to instructions, we would be replacing one mystery for another. The mystery of how mindless-Bob discovered a familial connection would be replaced by the mystery of how he discovered a familial connection by playing around with a diagram with distinctive shapes. Hence, there is more to understanding this process than simply describing the syntactic features of the diagram and how Bob responds to them.

This suggests a very different reading of the question we posed earlier. Instead of asking what features of the marks mindless-Bob uses to solve the problem, we can instead treat the question as asking what it is *about those* marks that, when used in that way, lead him to successfully perform the task in question. We are now asking *why* those markings eventually provide mindless-Bob with a solution when he uses them in accordance with the instructions. And the answer is that the marks on the paper do, in fact, accurately model the real-world family trees. Even when mindless-Bob fails to consciously interpret the marks on the paper, they are still serving as representations in a model that he has (unknowingly) built and is now exploiting. His scribblings on the paper help generate answers because those scribblings share a structural similarity to the relevant state of affairs he is investigating. We can't fully understand how mindless-Bob performs the operation of figuring out how two people are related unless we understand his operations as involving the implementation of a model. And to understand his operations as an implementation of a model, we need to look at the elements of these operations – in particular, the marks on the page – as representations of people and kinship relations.

In much the same manner, theories of the CCTC claim that to understand how the brain solves various cognitive tasks, we need to see it as implementing a model or simulation of the relevant target domain via a process of symbol manipulation. And to understand it as implementing a model via symbol manipulation, we need to treat the symbols themselves as representations of aspects of whatever is being modeled. Understanding how computers work involves understanding more than the nature of their physical operations. We also want to understand what it is about those specific operations that enable the system to perform some sort of task. We need to ask not only, "What is the causal/physical nature of this system?," but also, "What is it about the causal/physical nature of this system that enables it to solve this particular problem?" And the CCTC answer is this: These syntactic/physical operations are successful in solving this problem because they implement a model of a problem domain, and, as such, employ elements that *stand for* various aspects of that domain.[8] It is irrelevant that there are no components of the system that consciously interpret the symbols; that doesn't prevent the system from using some of its components as symbolic surrogates while running its simulations. The CCTC says that we should understand human cognition in exactly the same way. It claims that cognition should be understood as a process in which brains run simulations, and consequently employ representations of aspects of whatever domain is simulated.

Because the usual sense in which computational systems are said to do modeling is different than the sense I intend here, it is important to make the following distinction to avoid confusion. Computational systems and theories are themselves often regarded as providing models of brain processes. In this sense of cognitive modeling, the brain is the target of the modeling. But in the sense of modeling I am now speaking of, the brain is claimed by CCTC to be the "modeler," not the "modelee." That is, classical computational theories say that when the brain performs a given

[8] A number of people have suggested that I link the explanatory importance of S-representation too closely to the *success* of the cognitive system. But it is important to see that some degree of success is always presupposed in any specification of the cognitive task or capacity we are trying to explain. Without some success, we can't say that the cognitive operation we are trying to understand is actually instantiated by the system. After all, we don't say a rock does face recognition very, very poorly – we say it doesn't do face recognition at all. So the claim here is that one of the main explanatory goals of a cognitive theory is to explain how a given system (like the human brain) performs a cognitive task, and that requires assuming that it actually does perform that task, which in turn requires assuming that it performs it at least somewhat successfully. S-representation is needed for achieving this explanatory goal because it enables us to see how an internal structure is functioning as a model or simulation that enables certain systems to perform the operation in question.

cognitive task, the brain itself constructs a model of the relevant domain, and consequently uses representations of aspects of that domain as elements of the model. Cognitive models (in the usual sense) are models of what, according to the CCTC, is *itself* a modeling process. In effect, the computer model of the mind claims that the brain is actually doing what ordinary computers often do when they run simulations of various real-world processes.

Hence, it should be clear how, on this conception, brain states that are posited as part of a computational process (brain states that function as data structures) actually *serve as* representations in such a process. They do so by serving as constituent elements of a model or simulation that is exploited by the system when doing some cognitive task. In this context, "standing for" amounts to "standing in for," since problem-solving as model-building (or simulation-running) makes little sense without component elements of the model (or simulation) that function as representations. Haugeland captures the basic idea when he says, "[t]hat which stands in for something else in this way is a *representation*; that which it stands in for is its *content*; and its standing in for that content is *representing* it" (1991, p. 62). Or, to adopt Swoyer's language, computational systems (and, ex hypothesis, the brain) perform a type of mechanized surrogative reasoning. Surrogative reasoning requires surrogates, i.e., representations, and in computational accounts that job description goes to the symbolic data-structures. The content of the symbols is explanatorily relevant for their job because if the symbols don't stand for anything, the system in which they function can't itself serve as a model or simulation of the target domain, and we would have to abandon the central explanatory strategy offered by the CCTC. The job description challenge is successfully met with S-representation because we are provided with a unique role that is recognizably representational in nature and that fully warrants our saying the state serves a representational function. Moreover, this role of serving as a stand-in for some aspect of a target domain in a computational model or simulation is sufficiently distinctive and special to allow us to distinguish the representational elements of a system from the non-representational.

Besides answering the job description challenge, the functional role associated with S-representation allows us to account for other intuitive aspects of representation. For example, while a variety of factors may be necessary for a fully satisfactory account of S-representation content, it is clear that one significant factor will be a symbol's functional "place" in a model or simulation. If we ask how the tail section of a model plane in a wind tunnel comes to represent the tail section of a real plane – how the tail section of the real plane (and not the front nose section) comes to be the intentional object of the model's tail – the answer will appeal to the way the model is structured, how that structure leads to a certain kind of isomorphism with the real plane, and how that isomorphism maps one element of the model to an element of the target. Thus, S-representational content is linked to the sort of role the representation plays in the system's overall problem-solving strategy. In the CCTC-style explanations, the organization of the model and the nature of the resulting isomorphism with the target determines, in part, what it is that a given component of the model or simulation represents. Moreover, we get a sort of misrepresentation when the isomorphism breaks down. If the wingspan of the model tail wing is disproportionately longer than the wingspan of the actual tail, then that aspect of the model is in error and misrepresents the tail section of the real plane. Misrepresentation is a case of inaccurate replication of the target domain. Inaccurate replication occurs when and where the model or simulation fails to maintain its organizational (or structural) isomorphism with that which is being modeled.

In our discussion of the IO notion, we noted that there is a sort of mutual dependence between something having the function of serving as an IO representation, and a sub-system having the function of performing some internal computation. A similar sort of mutual dependence exists with S-representation. S-representations are necessary for a system's construction of a model *and* it is a state's participation in a model that makes it an S-representation. The constituent representations help make the embedding structure a model or simulation, and it is the embedding structure's status as a model or simulation that makes the constituent elements representations. This may initially appear to be a vicious circle. But remember that S-representation is a functional notion; hence, to do their job S-representations need to be embedded in the right sort of system. But such a system (i.e., a modeling or simulating system) isn't possible unless there are structures serving as S-representations. Thus, we have the same sort of mutual dependence that one often finds with functional kinds. Something must play a certain role in a system; but the relevant system itself isn't possible without components playing such a role. A person is a soldier only by virtue of being a member of an army. But there can be no army without members serving as soldiers. So soldiers and armies come together as a package. Similarly, something is an S-representation by virtue of the role it plays as part of a model of the target domain. But something needs to play that representational role (the role of standing for specific elements of the target) for there to actually be such a model. Models and simulations require S-representations to exist and nothing is an S-representation unless it functions as part of a model.

The explanatory value of S-representation becomes clearer if we consider how this notion offers an avenue of rebuttal to the anti-representational challenges posed by Searle and Stich. Since the case of mindless-Bob is simply a variant on the Chinese Room, we have already seen how the notion of S-representation comes into play under the sort of conditions Searle's argument exploits. Ex hypothesis, the room produces appropriate outputs in response to the inputs it receives. Thus, a computational account of the room would need to include an explanation of how it consistently does this. Syntactic symbol manipulations are only part of the story – we also want an explanation that tells us what it is about those manipulations that produces continued success (despite the ignorance and lack of understanding on the part of the manipulator). Depending on the details, one answer proposed by CCTC theories is that those manipulations model some target domain, and thus involve S-representations. In fact, if the program used by the Chinese Room is like the sort of architecture that inspired Searle's argument, such as Shank and Abelson's (1977) SAM (for "Script Applier Mechanism"), then its success would clearly involve models and hence S-representations. Shank and Abelson's theory uses "scripts," which are stored data structures that serve to symbolically replicate some state of affairs, like the central features of riding a bus or eating at a restaurant. As models of those activities, they allow the system to answer various questions which can be generated from the stored "background knowledge" in the script. If the symbols in the Chinese Room constitute scripts of this sort, then they serve as representations, not because there is some conscious interpreter who understands them as such, or because the people who designed the system intended them that way, but because the overall system succeeds by exploiting the organizational symmetry that exists between its internal states and some chunk of the real world.

Of course, the details of any specific account are less important than the core idea that classical theories invoke models and simulations and thereby invoke S-representations. Is S-representation comparable to full-blown conscious thoughts? No, it is a technical notion of representation based on our commonsense understanding of things like maps, invoked by a theory to explain cognition in a certain way. Searle is correct that the account of cognition offered by the CCTC fails to present a notion of representation that captures all aspects of our ordinary, commonsense understanding of thinking and thought. Nothing in the CCTC should lead us to conclude that Searle is wrong in asserting that the Chinese Room, as such, does not instantiate a full-blown mind. But the issue of whether or not full-blown minds could be instantiated by any system running the right program can be separated from the question of whether or not the CCTC provides a representational theory of how the brain works. What should *not* be conceded to Searle is the proposition that the CCTC fails to invoke *any* explanatorily valuable notion of representation. It should not be conceded that the only sense in which classical symbols serve as representations in computational processes is the artificial "as if" sense that is only metaphorical and has nothing to do with real representation. What the CCTC shows us is that a notion of representation can do explanatory work, *qua* representation, even in a purely mechanical problem-solving system.

Similarly, our earlier discussion revealed what Stich's purely syntactic account of computational processes would leave out. The question of why a system works is every bit as important as *how* it works. But a syntactic approach would largely ignore the former question. A syntactic story would reveal the process whereby the symbols come to be shuffled about in various ways. But it would not tell us what it is about those symbol shufflings that leads the system to consistently produce the appropriate responses. It would ignore the central aspect of the classical account that answers the question, "why do *these* syntactic operations enable the system to perform as well as it does?" A purely syntactic account would leave us blind to the fact that computational systems are doing surrogative reasoning because it would prevent us from seeing that computational structures serve as representational surrogates.

In fact, S-representation reveals a weakness in one of Stich's main arguments for the syntactic theory. Stich suggests scientific psychology should be guided by the "autonomy principle," which holds that "any differences between organisms which do not manifest themselves as differences in their current, internal, physical states ought to be ignored by a psychological theory" (1983, p. 164). In defense of this principle, he offers what he calls the "replacement argument." Since cognitive psychology is in the business of explaining the inner workings of the mind/brain, any historical or environmental dissimilarities between an original and physical duplicate that fail to generate actual causal/physical dissimilarities should be ignored by cognitive (and, by extension, computational) psychology. If a robot on an assembly line is replaced with an identical duplicate, then, Stich argues, cognitive psychology should treat the original and the double as the same. The same goes for human agents. But since the content of the agent's belief-type states *is* based upon historical or environmental factors, then a psychology that pays attention to content will treat the human duplicate as different from the original, thereby violating the autonomy principle. Given the intuitive plausibility of the autonomy principle, Stich takes this to show that computational psychology ought to *ignore* content and drop the notion of representation altogether.

The problem with this analysis is that it is based upon a conception of representation in the CCTC that is too narrow. Stich's argument at least tacitly adopts the Standard Interpretation and thereby treats computational representations as analogues for propositional attitudes. For propositional attitudes, content is arguably due entirely to external, causal-historical facts, and thus physically identical systems may differ in terms of folk mental representations. By contrast, S-representation carries the possibility of understanding content in a way that is far less dependent upon causal-historical details. Above I've suggested that S-representation stems from the use of an inner model in some cognitive task that is properly isomorphic with its target. Because any replica placed in the same situation will (by virtue of being a replica) employ internal structures functioning in the same manner as the original, it is also presumably using the same sort of model and thus the same sort of S-representations. If, say, a robot that is using an inner map to successfully maneuver in some environment is replaced by a physically identical system, then the explanation of how the duplicate performs the same task would also need to appeal to the same sort of inner map. The use of a map is not so directly dependent upon the history or source of the map, and it would be bizarre to claim that the performance of the original robot involves inner elements that cannot be invoked when we explain the success of the physically indistinguishable duplicate in the same situation. So, intuitively, the same S-representational account would apply to the replacement that applied to the original. Thus, S-representation passes the replacement test and thereby satisfies the autonomy principle. If the replacement argument is intended to show how the syntactic approach to the CCTC is superior to a representational approach, then it fails to do so once we appreciate the importance of S-representation to CCTC explanations of cognitive processes.

By suggesting that there is a theory-based notion of representation that is built into the explanatory framework of the CCTC and that accords with the autonomy principle, I do not mean to suggest that all external, environmental factors are completely irrelevant. In the case of the duplicate robot, I'm suggesting its *use* of a model (and thus S-representations) depends upon its performance of a specific task, and its performance of a specific task depends upon the circumstances in which the system is embedded – on the problem-solving environment. I'm suggesting that while the causal history of a map or model is (perhaps) irrelevant to its current usage, the specific environment in which it is employed may be highly relevant to questions about content. In the next section, I'll argue that the task environment helps determine how a posited model is used, which in turn determines, in part, what S-representations represent. In short, the task-environment a model is plugged into helps determine the model's target, and the model's target helps determine the content of S-representations. This arrangement suggests a possible solution to a traditional problem associated with S-representation.

This barely scratches the surface of all of the different dimensions and worries connected to S-representational content,[9] some of which will be further addressed below in sections 3.3 and 3.4. As with the interior IO notion of representation, the S-representation notion requires a more detailed and sophisticated elaboration than I can provide here. But remember my aim is to only show that there is a notion of representation at work in CCTC theories that answers the job description challenge by describing structures playing a functional role that is both naturalistic and recognizably representational in nature. Both the IO notion and the S-representation notion do this. And yet both notions are a bit like the Rodney Dangerfields of representational posits – they get no respect, or at least not as much respect as they deserve. Much of this is because of two traditional problems that are often assumed to undermine their theoretical value. Since I think these problems are overblown, it will help to look at them more closely.

3.3 TWO OBJECTIONS AND THEIR REPLIES

All notions of representation have their difficulties, and the two notions which I have argued are central to the CCTC explanations are no exception. In this section I would like to address what I take to be the two most common criticisms of these notions. One charges that mere isomorphism does not provide a sufficiently determinate sort of representational content, and thus the notion of S-representation is critically flawed. The other criticism suggests that these representational notions are really too weak and make representation nothing more than a heuristic device. My aim will be to demonstrate that a better understanding of the explanatory work that

[9] For a much fuller discussion of these issues, see Cummins (1989, 1996) and Horgan (1994).

these notions are doing in the CCTC reveals that the criticisms are much less damaging than they initially seem.

3.3.1 Challenge 1: indeterminacy in S-representation (and IO-representation) content

A problem for S-representation in particular (but also could be developed into a complaint about IO-representation) is one that we've already briefly touched on. As we've seen, the notion of S-representation is based upon some sort of isomorphism between the model or simulation and the target being modeled or simulated. But, notoriously, isomorphisms are cheap – any given system is isomorphic with many other systems. For instance, in the non-mental realm, maps provide an example of S-representation whereby the individual figures and lines on the map stand for specific aspects of an environment by virtue of the overall isomorphism between the map and that environment. But in truth, the collection of lines and figures on a simple map can be equally isomorphic with a range of different environments and geographic locations. Hence, which parts of the world they *really* designate cannot be determined by appealing to isomorphism alone. Or returning to the computational realm, Fodor and others have emphasized that two systems simulating different events – one, say, the Six Day War, the other a chess game – could be functionally identical, so that "the internal career of a machine running one program would be identical, step by step, to that of a machine running the other" (1981, p. 207). Consequently, if presented with such a program, there would be no fact about whether the inner symbols S-represent the Sinai Peninsula and Egyptian tanks, or chess board locations and pawns and rooks. So it looks like the S-representation notion has a serious problem of content indeterminacy. Something is a representation by participating as part of a simulation or model, which in turn is a simulation or model *of* that aspect of the world with which it is isomorphic. But models and maps are isomorphic with many different aspects of the world, so the representation is potentially about a wide array of things. Simply mirroring the organizational configuration of some state of affairs is not sufficient to make something a model or simulation *of* that state of affairs, even if we limit the amount of complexity we build into our interpretation scheme. The target of any model is inherently indeterminate, and thus the content of any element of that model is indeterminate. But actual representations have determinate content, so being an element of such a model or simulation is not enough to make something a representation.

Yet I believe this verdict is too quick. For those invoking representations as part of the CCTC, the indeterminism worry is, I believe, a red herring that stems from a failure to appreciate the nature of the explanatory challenge facing the cognitive theorist. The explanatory challenge is typically not to explain *what* a cognitive system is doing, but *how* it is doing it. Competing theories of cognition, including the classical paradigm, offer hypotheses about inner processes that are designed to account for our various mental abilities. These abilities are thus taken as a given, as a starting point for our theorizing. The explanatory challenge can be characterized in this way: "Given that we successfully do such and such (recognize faces, maneuver through a complex environment, determine the grammaticality of sentences, etc.), provide an explanation of how our brains do it." So the CCTC starts with a specified cognitive ability, and then offers an explication of how that ability is realized in the brain. In the process, the CCTC posits S-representations that play a certain sort of functional role. In the case of mindless Bob, that role was instrumental in helping him to answer various questions about a given family, despite Bob's ignorance of what he was actually doing. Even though Bob isn't interpreting his model, it still makes perfectly good sense to explain his success by saying he is using a model, and moreover, a model of one particular family. It doesn't really matter that the drawing is isomorphic with dozens of other family trees or, for that matter, dozens of other states of affairs. It is used as a model of *this* family, and not some other, because this is the family that is the subject of Bob's problem-solving efforts. In other words, the problem domain itself – the situation that Bob finds himself in, and for which he (mindlessly) employs his diagram – is what makes determinate the target of his model, and thus the content of the representations that make up the model. The indeterminacy problem arises when there is no way to fix the target of a model. In cognitive explanations, however, the explanandum itself typically *does* fix the pertinent target and thereby determines what it is that is being modeled. Looked at another way, if the brain is indeed performing a specific cognitive task, then a classical computationalist gets to posit representations with determinate content because she gets to posit models and simulations used in the performance of *that* task (and not some other). If the true explanation of how my brain succeeds in some mental task is that it relies on a model, it simply doesn't matter that, taken as an abstract structure, the model is isomorphic with other things. It may be, but that doesn't undermine the way my brain is using it here.

Of course, we've now traded one sort of problem for another. The original problem was determining what a simulation or model is a simulation or model of. I'm claiming this can be settled by looking at how the model is actually being used in specific problem-solving situations – by looking at the actual task for which the structure is used. But that just shifts the problem of determining the target of the model to one of determining the exact nature of the task the system is performing. What warrants my saying that Bob is constructing answers to questions about a family, and not doing something else?

Well, lots of things. First, it is important to see that this is an issue that we are going to need to address no matter what our theory is. Specifying the nature of the cognitive task a system is performing, while a deep and thorny topic, is a deep and thorny topic that everyone needs to confront – it is not a problem that is unique to accounts that appeal to S-representation. Second, there are various promising strategies for addressing this issue. The most obvious builds on the fact that representational systems do not operate in a vacuum; they are embedded in the world in various ways, and this embedding helps determine the actual tasks they are attempting to perform. Consider ordinary psychological explananda. There is no deep mystery in trying to decipher the task a rat is performing as it attempts to make its way through a maze for some food. It is trying to navigate its way through a maze. Which maze? The one it actually finds itself in. If a theory says the rat is using an internally stored map for navigation, then the map is, by virtue of that very use, a structure that is used to model this particular maze. It simply doesn't matter for our explanatory purposes that there are lots of other mazes or terrains in the world that the map is isomorphic to. So, by looking at the way representational systems are embedded in the world, we can specify the particular tasks they are confronting. And by specifying the particular tasks they are confronting, we can specify the particular target of whatever models or maps the system is using for that task. And by specifying the targets of whatever models or maps that are being used, we can specify what it is that the elements of those models and maps actually stand for. A model's constituent parts stand for those things they are used to stand *in* for during actual episodes of surrogative reasoning.

In his more recent writings, Cummins (1996) has suggested that physical structures represent all things with which they are isomorphic and, thus, representations never have a fixed, determinate content. The way I suggest we understand S-representation is quite different. Parts of a model don't automatically represent aspects of some target *simply* because the model is isomorphic to the target. Rather, components of the model *become* representations when the isomorphism is exploited in the execution of surrogative problem-solving. The CCTC claims that the brain employs representations because the brain uses some sort of a model of the target (or some aspect of the target) and neural states serve as representational parts of that model. Yet it's possible the same basic computational model or simulation could be isomorphic with some other target, and therefore could be used in the execution of some other cognitive task, in some other problem-solving situation. But that possibility doesn't matter because that's not what *this* brain is using it for *now*. The neural symbols really do stand for, let's say, board positions of a chess game, and not the positions of armies in the Sinai Peninsula, because what the theory is about is a cognitive agent playing chess and not fighting a war. Given that the agent is playing chess, classical computationalists can say he is doing so by running simulations of possible moves – simulations that are comprised of representations. A cognitive agent is figuring out chess moves and not battle strategy for the Six Day War because the agent is causally linked to a chess game and not a battlefield.[10] Thus, the content of S-representation can be fixed by the target of the model, and the target of the model is fixed by the cognitive activity we want explained. The cognitive activity we want explained, moreover, is typically dependent upon the way the system is currently and causally engaged in the world. The upshot is that the content indeterminacy problem is simply not as big a challenge for S-representation as is generally assumed.[11]

3.3.2 Challenge 2: IO-representation and S-representation aren't sufficiently real

Throughout this discussion I've made repeated appeals to the *explanatory benefit* of positing representations, or to the *explanatory pay-off* of invoking IO and S notions of representation. Moreover, I've defended both notions by insisting that there actually is such a pay-off. In the case of the

[10] Of course, the same sort of causal considerations could *not* help us define mathematical cognitive processes or different forms of abstract or hypothetical reasoning in which we have no clear causal connection to the target domain.

[11] In my analysis, I've deliberately avoided appealing to historical factors or to the way the cognitive map or model is constructed when specifying the map or model's target. This is because I believe that what a map or model targets is intuitively more dependent upon how it is used than on where it came from. I believe, say, that a map that is used to navigate a particular terrain is serving as a map *of* that terrain, even if it was originally written as a map of some other environment. But others may find this implausible and believe that diachronic considerations are indeed key to determining a map or model's target. This would provide another strategy for handling the indeterminacy problem, and would support my main point that the problem can indeed be handled.

IO-representation, the notion allows us to employ an explanatory strategy of functional analysis whereby the inner sub-systems can be seen to perform tasks germane to the larger explanandum. In the case of S-representation, the notion allows us to treat the system as employing a model or simulation, which in turn helps us to explain how it succeeds in performing a given cognitive task. However, this emphasis upon the explanatory role of representations has a down side. It suggests that a structure's status as a representation is entirely dependent upon our explanatory interests and goals – that things are representations only insofar as we gain some explanatory benefit from treating them that way. This implies that the notions of representation under consideration serve as something like heuristics or useful fictions, similar to a frame of reference, or the proverbial family with 1.5 children. As such, they don't correspond to anything objectively real in nature. IO- and S-representations would exist only to the extent that we take a certain explanatory stance toward a system, and without the projection of our intentional gloss, there would be no such sorts of things.[12]

The philosopher who has been the strongest advocate of the view that the having of representations depends on our trying to explain or predict the behavior of the system is Daniel Dennett (1978, 1987). In our efforts to understand any given complex (or even simple) system, there are, according to Dennett, different explanatory stances or strategies that we can adopt. First, we can adopt the "physical stance" and use an understanding of the physical inner workings of the system to explain and predict how it responds to different inputs. Or, alternatively, we can adopt the "design stance" and predict behavior by using what we know about the sort of tasks the system was designed to perform. Finally, we can sometimes adopt what he calls the "intentional stance." The intentional stance involves treating a system as a rational agent with beliefs, desires and other folk representational states. Dennett has argued extensively that being a "true believer" is little more than being a system whose behavior can be successfully explained and predicted through the ascription of beliefs and other propositional attitudes. If we can gain something by treating a Coke machine as having, say, the thought that it has not yet received enough money, then the Coke machine really does have such a thought.

[12] Michael Devitt has put it to me this way: "It is as if you don't think representation is a REAL PROPERTY of anything; it's just a matter of when it's appropriate to take the 'intentional stance' toward it" (personal correspondence).

Despite his extensive and often inventive arguments for this perspective on intentional states, few philosophers or cognitive scientists have adopted Dennett's interpretationalist account. For any given system, Dennett closely ties the possession of mental representations to the explanatory activities of other cognitive agents – on the sort of explanatory perspective they adopt. Yet most people think these are the wrong *type* of considerations for identifying representational systems because mental representations are regarded as objectively real, observer-independent aspects of the world. The criteria for being a representation should be of the same nature as the criteria for being a kidney or a virus; namely, actual intrinsic or relational properties that are either present or not, regardless of how the system could be explained by others. Returning, then, to my analysis of computational representations, the complaint is that the notions we've explored in this chapter are overly "Dennettian." It appears that, like Dennett, I've offered an analysis whereby the positing of representations stems from the explanatory strategies and goals of cognitive scientists who need them to adopt certain perspectives on a proposed system's operations. It seems these notions of representation serve as representations not for the system, but for the psychologists attempting to understand the system in a certain way. Yet there is surely more to representation than that. If the CCTC is a framework that invokes *real* representations, then it needs to do so in a way that is far less observer-dependent and far more objectively real than I've suggested here.

My response to this challenge is that the notions of representation we've looked at here *are* fully objective and observer-independent, and any appearance to the contrary is simply an artifact of my emphasis upon the explanatory role of representations in science, and not a deep fact about their metaphysical status. Indeed, when we step back and look at how theories in the CCTC invoke representational states, we can see they have the same status as other scientific posits whose existence is assumed to be objectively real.

To begin with, it should be acknowledged by all that there is a very weak sense in which most of the things we deal with in our daily lives are observer-dependent. The sense I have in mind is just this: It is at least theoretically possible to view any system as nothing more that a cloud of interacting molecules or even atoms; hence, our *not* viewing the system in that way requires our adopting a certain stance. In this weak and uninteresting sense of "stance dependence," any notion that is not part of the vocabulary of basic physics can be treated as unreal or "merely heuristic." Yet it shouldn't bother the advocate of the notions of representations presented here if it should turn out that the representations are observer-dependent in *this* sense. In other words, if the argument that representations are observer-dependent is simply that it is *possible* for us to view a computational system as nothing more than a bunch of interacting molecules (that is, that it is possible to adopt the physical stance), then this way of being observer-dependent shouldn't cause us concern. If IO and S-representations are unreal only in the sense in which trees, minerals, hearts, mountains and species are unreal, then a realist about representation should be able to live with *that* sort of "anti-realism."

We can therefore assume that the anti-realism challenge alleges that the IO and S notions of representation are observer-dependent in a stronger sense than this. But it is much harder to see how a stronger sense of observer-dependence applies to these notions. Consider again the way in which the IO and S notions of representation are invoked. The CCTC is a general account of how the brain works; more specifically, it is a theory about how the brain performs various cognitive tasks. It claims that the brain does this by performing symbol manipulations of a certain sort. In many (or even most) versions of the theory, the symbol manipulations attributed to the brain are those that involve a) inner sub-systems that perform various sub-tasks that are stages of the task being explained and/or, b) models or simulations of the relevant target domain. Both of these types of operations require representational states. The sub-systems employ representations because the sub-tasks convert representations relevant to one aspect of the cognitive tasks into representations of another aspect. And the models employ representations because the components of all models stand in for different elements of whatever is being modeled. So, if the CCTC is the correct theory of how the brain works, then the brain really uses inner representations.

Now it is not at all clear where in this account an anti-realist or observer-dependent interpretation of representation is supposed to arise. While it is true that the account links the having of representations to other things the system is doing (namely, using inner sub-systems and models), it is unclear how this alone is supposed to make representations useful fictions. There are, it seems, only two possible strategies for arguing that the CCTC leads to a sort of anti-realism. The first would be to drive a wedge between the status of the representation and the sorts of processes that the CCTC invokes to explain cognition. On this scheme, one might claim that inner sub-systems and models don't actually require representational states; hence, if we treat structures as representations, we are simply adopting the intentional stance for heuristic or instrumental reasons. The second way would be to concede that sub-systems and models require representations, but to then argue that the sub-systems and models themselves are also observer-dependent and therefore not sufficiently real. Yet neither of these strategies is terribly compelling.

With the first strategy, it is difficult to see how the argument could even get started. There is no discernible way that something could serve as, say, an adder, without it also being the case that it converts representations of numbers into representations of sums. Without such a conversion of representations, it simply wouldn't be doing addition. So too for countless other tasks that inner sub-systems are routinely characterized as performing. Along similar lines, it is hard to see how anything could employ a model or simulation of some target domain, and yet, at the same time, not have it be the case that the individual elements of the model or simulation stand for aspects or features of the target. If we are committed to the reality of the kind of processes classical computationalists claim are going on in the brain, then we are committed to neural structures really serving as inner representations. Of course, one could argue that brains don't actually implement the sorts of processes described by the CCTC. But that would be to argue that the CCTC is false – not that it employs an observer-dependent notion of representation.

The second strategy is to allow that representations are as real as the processes in which they are embedded, but to then argue that those processes themselves are useful fictions, interpretation-dependent, or subjective in some similar sense. With this view, the brain is not really employing inner sub-systems that perform computations which are stages of larger cognitive tasks; or, alternatively, the brain is not using models or simulations in any sort of objective sense. These are just subjective interpretations of physical processes. Indeed, this challenge could be seen as a more general worry about the very nature of computational processes or simulations themselves. The writer who is best known for this sort of criticism of computation is, once again, Searle (1990, 1991). Searle originally allowed that the Chinese Room, though lacking symbols with real meaning, was at least performing syntactic operations on formal tokens. He has since reconsidered this matter and now holds that computational systems and syntactic processes also fail to exist in any robust, objective sense. He tells us,

> Computational states are not *discovered within* physics, they are *assigned* to the physics . . . There is no way you could discover that something is intrinsically a digital computer because the characterization of it as a digital computer is always relative to an observer who assigns a syntactical interpretation to the purely physical features of the system . . . to say that something is *functioning as* a computational process is to say something more than that a pattern of physical events is

occurring. It requires the assignment of a computational interpretation by some agent. (1990, pp. 27–28)

Unlike the Chinese Room argument, Searle's argument for the position that computational processes are observer-dependent is somewhat hard to discern. In fact, Searle's discussion appears to provide us with less in the way of an argument and more in the way of alternative characterizations of his conclusions. In spots, when Searle tells us that "syntax is not intrinsic to physics" (1990, p. 26) and "syntax is not the name of a physical feature like mass or gravity" (1990, p. 27) it sounds as though he is defending the uninteresting view discussed above, that anything that is not described using the terminology of basic physics is observer-dependent. In other spots, Searle muddies the water by lumping together user-dependence and observer-dependence. Yet if the brain uses computational programs in the same sense in which, say, our body uses an immune system, this notion of use would be fully objective (after all, chairs may be sitter-dependent, but this doesn't make chairs observer-dependent). At one point, Searle tells us that "on the standard definitions of computation, computational features are observer relative" (1991, p. 212). But he doesn't tell us exactly how the "standard definitions of computation" lead to the view that differences in computational systems are in the eye of the beholder. There really is no generally accepted principle of computation that would rule out the possibility of distinguishing computational systems or programs by appealing to their causal/physical architecture, or that would entail that all computational processes are objectively indistinguishable, or that would suggest that there is no observer-independent sense in which my laptop is using Wordstar but the wall behind me is not. So on the one hand, there is a sense of "observer-dependent" that applies to computational systems and processes. But it is a completely uninteresting sense in which virtually everything is observer-dependent. On the other hand, there is a more interesting sense in which things might be observer-dependent – like being a funny joke. But as far as I can tell, Searle hasn't given us an argument that programs are observer-dependent in *that* sense.[13]

[13] Searle's view is also puzzling given his own claims about the value of weak AI. Since the sort of program a system would be running would be a matter of interpretation, there wouldn't be any objective fact about the quality or even the type of program implemented on a machine. Any problems that arose could be attributed not to the program, but to our perspective, so the difference between a good computer simulation of a hurricane and a bad one, or, for that matter, between a simulation of a hurricane and a simulation of language processing would all be in the eye of the beholder. It is hard to see how such an observer-dependent set-up could serve to *inform* us about the actual nature of various phenomena.

Since my goal is not to defend the CCTC but to instead defend the idea that the CCTC makes use of valuable and real notions of representation, perhaps the appropriate way to handle this worry is as follows. If you want to claim that the notions of representation discussed in this chapter are observer-dependent fictions, then you must do so at a very high cost. You must also adopt the view that computational processes themselves are observer-dependent, and this has a number of counter-intuitive consequences. For example, such an outlook would imply that a pocket calculator is *not really* an adder, computers *don't actually* run models of climate change, and no one has ever *truly* played chess with a computer. It would imply that whether or not your lap-top is infected with a virus or running a certain version of Windows is simply a matter of your perspective – that the distinction between different programs is no more objective than the distinction between images seen in ink-blots. Moreover – and this is the key point – you would be denying that we could discover that, at a certain level of analysis, the brain really does implement symbolic processes of the sort described in the CCTC, particularly those that appeal to inner sub-systems and models, and does so in a way that it might not have. Since this strikes me as a radical and counter-intuitive perspective on the nature of computation, the burden of proof is on someone to make it believable. As far as I can see, this hasn't been done.

3.4 CCTC REPRESENTATION: FURTHER ISSUES

No doubt many will find this analysis of CCTC notions of representation incomplete, which, in many respects, it is. But recall that our goal has been fairly modest. It has not been to provide a complete theory of representation that solves all problems associated with a naturalistic account of representation. Rather, it has been to defend the idea that the CCTC posits representations with considerable justification. It has been to show that IO-representation and S-representation are sufficiently plausible, robust, and explanatorily valuable notions to warrant the claim that, contrary to what the Standard Interpretation might lead one to think, the CCTC is indeed committed to internal representations. Nonetheless, despite these limited goals, there are further issues associated with these concepts of representation that warrant further attention.

3.4.1 Is IO-representation distinct from S-representation?

While Cummins's 1989 account of computational representation has served as the basis for much of the analysis provided here, my account departs from his in two important respects. First, Cummins treats the *exterior* IO notion of representation as the central notion at work in the CCTC, and fails to say much about the interior IO notion. Moreover, Cummins argues that the same notion is at work in connectionist accounts of cognition, as networks also convert representational inputs into representational outputs (1989, 1991). According to this view the CCTC and connectionist theories actually employ the same notion of representation. I believe this is a mistake. The error stems from treating exterior IO-representations as *part of* either theory's explanatory machinery. As I argued above, the exterior notion generally serves to frame the *explanandum* of cognitive science. That is, the typical cognitive task we ask a theory to explain is characterized as a function whereby input representations of one sort are converted into output representations of another sort. Thus, these exterior representations are not so much a part of the explanatory theory as they are a part of the phenomenon we want explained. This is not the case for the notion of *interior* IO-representations. These actually do form part of the distinctive explanatory machinery of the CCTC because classical theories often explain cognition by appealing to a hierarchically organized flow-chart. Since the internal sub-routines require their own inputs and outputs to be treated as representations, the interior IO notion becomes a notion of internal representation that is, by and large, unique to the CCTC.[14]

The second area where my analysis has departed from Cummins's original treatment concerns my distinction between two sorts of notions of representation at work in the CCTC. Whereas Cummins appears to treat the inputs and outputs of computational processes as S-representations, I've chosen to separate these as two distinct notions doing different explanatory jobs. It is fair to ask if this is the right way to look at things: if there really are two separate notions at work, as opposed to just one.

The reason I distinguish IO-representations from S-representation is because I am, recall, demarcating notions of representation in terms of the sort of explanatory work they do. Putting this another way, I am distinguishing notions of representation in terms of the way they actually *serve as* representations according to the theory in which they are posited. My position is that the way in which a structure serves as an interior IO-representation is different from the way it serves as an S-representation. In the case of the former, the job is linked to an internal sub-module or

[14] Of course, there are a number of elaborate connectionist networks that also invoke inner subsystems, and thus also employ the interior IO notion. Yet as I've noted in other works (Ramsey 1997), this is not the standard notion of representation that appears in connectionist modeling.

processor performing computations relevant to the overall capacity being explained. Such an inner sub-system typically receives representations as inputs and generates representations as outputs, so that is how representation comes into the explanatory picture. In the case of S-representation, the story is quite different. There the job of representing is linked to the implementation of a model or simulation, which requires components that stand for the relevant aspects of the target domain. Thus, the explanatory appeal to representation is quite different than it is with the IO notion; in fact, it is quite possible to have one without the other. For example, we could have a cognitive system that is explained with a task-decompositional analysis invoking inner sub-systems transforming IO-representations, but that makes no use of a model or simulation (an organized cluster of connectionist networks might be such a system). Or, alternatively, there could be theories that use inner models or simulations of some target domain but that don't appeal to inner sub-systems that require representational inputs and outputs (some simple production systems might have this feature). Consequently, the two notions of representation are distinct and should be treated as such.

Of course, this is not to say that computational structures never play *both* representational roles. In CCTC accounts, data structures can serve as both IO-representations and S-representations. This might happen whenever the simulation involves sub-computational systems that serve as stages or segments of the model or simulation. For instance, in our original multiplication example, the internal addition sub-system would be part of a simulation of a mathematical algorithm in which numbers are multiplied via addition. The data structures generated by the adder represents sums *both* because they are produced by an adder and because, as such, they are part of the simulation of a mathematical process (a type of multiplication) that involves sums. In fact, many would claim that *all* numerical computations are simulations of various mathematical functions. If this is true, then the IO notion could be reduced to a special type of S-representation for these types of computational operations. It wouldn't follow that all IO-representations are special cases of S-representation, or that there aren't really two different explanatory roles associated with these two notions. However, I would not be surprised if a more detailed analysis revealed that a large number of CCTC models posited representational structures that do double duty in this way.

3.4.2 Cummins's abandonment of S-representation

As we noted, the idea that the CCTC framework typically invokes S-representations is at least partly due to the analysis provided by Cummins, as presented in his 1989 book *Meaning and Mental Representation*. In more recent writings, however, Cummins appears to claim that this notion of representation, at least as originally presented, is severely flawed (Cummins, 1996). While there is much that could be said on this topic, I briefly want to consider Cummins's reasons for rejecting S-representation as an account of computational representation to see if his new position undermines our analysis.

In his 1996 book, *Representations, Targets and Attitudes*, Cummins develops an account of representation that puts the problem of error at center stage. His account of error dwells on the possible mismatch between the actual content of a representation and what he refers to as its "application" to an intended target. To illustrate this, Cummins exploits a common type of classical computer architecture in which symbolic variables take specific values. Suppose the system is a chess-playing program with sub-systems that generate board states corresponding to actual elements of the game (these Cummins refers to as "intenders"). Suppose further that one such sub-system generates a slot that is supposed to be filled with a representation of the next board configuration, which happens to be P2. P2 is thus the target for representation. If all goes well, the slot will be filled with a representation of P2, i.e., RP2. This slot-filling (variable-binding) is what Cummins calls the application of the representation. Now, suppose the slot is instead filled with a representation of a different board position, namely, P3. An error would thereby occur because the intended target (P2) would not be represented by the representation that is applied (RP3). This sort of error is possible only when there is a mismatch between representation and target. Error is thus a form of *mis-application* of a representation with a fixed content to the wrong target. Because the content of the representation itself has no truth-value (it represents only the board position, not its status) the representation itself can't be false. The application, however, *does* have propositional content – in this case, it represents something like "The next board configuration will be P3." Since the next board position is P2, the content of the *application* is what is false.

Initially, Cummins's discussion appears to be only an extension (or perhaps special application) of his earlier account of S-representation. After all, the sort of computational account he invokes while describing representational error looks just like a computational account that uses S-representations as part of a simulation of a chess game. But Cummins explicitly rejects S-representation because, he claims, S-representational content cannot account for the sort of error just described. The content of S-representation depends on how the representation is used by the system. But all use-based accounts of content, he argues, identify representational content with the intended target, thereby making it impossible for misapplication to occur. Cummins's reasoning runs as follows: Use-based accounts of representational content make the content of the representation a function of how the representation is used by the system. But use amounts to the same thing as intended application. Any theory that makes content a function of how the representation is used claims that content is determined by its target application, so what the representation actually means must correspond with what it is intended (or applied) to mean. Hence, the content of the representation will always correspond with the intended target; hence, the two can never pull apart; hence, there can be no error. But error is something any serious theory of representation must explain, so use-based theories of representation don't work. The S-representation notion is also use-based, so it too doesn't work. What is needed is an account that makes content an intrinsic feature of the representation, something that is independent of how it is employed. For Cummins, a picture-based account of representation provides this, since the structural properties that make a picture a representation of something are intrinsic features.

For those of us impressed with Cummins's original account of how the notion of representation is employed in the CCTC, this newer position is a bit confusing. On the one hand, he appeals to a familiar sort of computational process (role-filling) to attack what appears to be a natural way to think about computational representation that he once endorsed. On the other hand, he rejects an account of representation based on isomorphic relations to targets, but he endorses an account of representation that also appeals to a form of isomorphism. Unfortunately, it would take us too far afield to provide a completely detailed analysis of this apparent change of heart. Instead, I'll offer a not-so-detailed analysis, suggesting that Cummins's newer account is mistaken about one key issue.

The crux of Cummins's argument is the idea that use-based accounts of content cannot drive a wedge between a representation's target and its content. But why should we think this? All that is needed is a way to distinguish between what the system needs or intends to represent on the one hand, and what the system actually represents on the other hand. Contrary to what Cummins suggests, accounts in which the content is based upon the representation's use have little trouble doing this. One way is to tell a story whereby the system intends to token a structure that, given how it is used, represents X, but accidently tokens a structure that, given how *it* is used by the system, represents Y. It needn't be the case that the intended causal role is the same as the actual (content bestowing) causal role for any given representational structure. That is, it needn't be the case that with a use-based account, a symbol slotted into the "Next Move" variable would automatically stand for the next move which, in the scenario described, would be P2. Instead, for sophisticated use-based accounts, a symbol would retain its content in such an application because it would retain a distinctive role in that application. When plugged into such a slot, a symbol would have a distinctive effect on the system, and this distinctive effect would contribute to its content *and* in certain situations, give rise to error. For example, the symbol RP3 would cause the system to respond differently in the "Next Move" application than the symbol RP2. The different effects of these symbols when applied to the same application contribute to their having different representational content. RP3, when put into the "Next Move" slot, has the sort of effects that are appropriate if in fact the next move is going to be P3. But since the next move isn't P3, this is a case of error. Since the next move is actually P2, the system needed to use RP2 to fill the "Next Move" variable because RP2 generates the correct simulation.

Part of what makes the tail-section of a model plane in a wind tunnel stand for the tail-section of a real plane is the role this segment plays in an explanatory and predictive model. But that doesn't prevent the possibility that the dimensions of the model's tail-section are in error, given the actual dimensions of the real plane's tail. So too, computational models can have faulty settings. The key point, then, is that you don't give up on error just because you think differences in computational roles also contribute to differences in the content of computational symbols. As noted above, it is far from clear that S-representational content is *only* determined by internal use – embeddedness can also intuitively contribute to content. But even if S-representation content should prove to be *entirely* a matter of use, Cummins's more recent analysis fails to give us good grounds for rejecting S-representation. Given the ways in which content and target can come apart even with a use-based account of content, it is still possible for S-representations to be mis-applied and thus it is still possible for S-representations to lead to the sort of error Cummins cares about.

3.4.3 What about rules?

There is yet another notion of representation traditionally associated with the CCTC that we have not yet explored in detail but that needs to be discussed. Classical systems are often characterized as employing a "rules and representations" architecture. This characterization is misleading insofar as it suggests that computational rules are somehow different from representations. Since computational rules are generally viewed as standing for various things, like discrete commands or instructions that serve to guide the system's operations, rules clearly are meant to serve as a *type* of representation. Indeed, the rules are often said to encode a computational system's explicit "know-how." The core idea is that the system performs its various internal processes by "consulting" symbols that encode instructions pertaining to specific operations. In other words, the architecture of the system is designed so that various causal transitions are mediated by, or indeed, "guided by," these representations. While the simulation notion of representation can be seen as the computational equivalent of a road map representing the relevant terrain, the explicit rule notion is the computational equivalent of traffic signs directing traffic.

With regard to rules, the central questions we need to ask are these: Are representations of rules a distinctive type of representation? If so, what type of explanatory work do they do? If not, can they be subsumed under the heading of interior IO-representation or S-representation, or should we instead just stop treating them as representations altogether? My position is mixed. Some types of rule representations are just a special case of S-representation, and thereby have real explanatory value. There are, however, some structures characterized as rule representations that cannot be treated as a type of S-representation or interior IO-representation. In these cases, I will argue, the structures are not actually serving as representations of rules at all.

We have been demarcating notions of representation by appealing to their alleged explanatory role. In the case of rules, that explanatory role is intimately connected to the sort of command the rule is thought to encode – what it is "telling" the system to do. However, computational structures are sometimes characterized as "rules" even though their content doesn't actually tell the system to *do* anything. For example, computational rules can encode conditionals, where both the antecedent and consequence of the conditional designate aspects of the target domain. Suppose the symbolic structure encodes an entailment relation like, "If condition X obtains, then state Y will come about." Generally, such a representation will be a component of some larger model of a target set of conditions that includes conditions X and Y. When this occurs, it is clear that such a representation is just a special form of S-representation. These counterfactual representations designate actual entailment relations and so make up an important element of a model, even if the antecedent does not obtain in the actual world. So in classical computational systems, conditional statements of this sort that are referred to as "rules" are just a special case of S-representation.

More often, though, a significantly different sort of conditional statement is assumed to be encoded by a rule. Here the antecedent still designates a real-world condition, but the consequence is thought to designate a real-world course of action. In other words, computational rules are often thought to have the content, "If condition X obtains, then *do* Y." When this happens – when the consequence is something like, "pick up the square block" or "move the Bishop to position Y" – we have what looks like a completely different sort of representation, one that is *prescriptive* as opposed to merely *descriptive*. Thus, it is less clear that this sort of rule qualifies as a special case of S-representation.

Still, I think a strong case can be made for treating prescriptive rules of this sort as a case of either interior IO-representation or S-representation. On the one hand, a computational sub-process may have as its output a representation of some conditional action rule. That is, there may be a sub-routine in a system that is designed to generate different strategies for responding to certain conditions. If so, then to regard this sub-system as having this function, we need to view its outputs as representations of the sort, "If X, then do Y." This would clearly be an instance of IO-representation. On the other hand, sometimes it may be more appropriate to regard the command as representing a stage of some real-world process being simulated, *even though* the computational system (or its real-world extension) is itself causally responsible for that particular aspect of the process being modeled. There is no obvious reason to claim that a computational system cannot, itself, participate in some of the transactions that comprise the target of its own simulations or models. There is no reason why a model user can't bring about some of the events that are part of what is being modeled. Given this, these sorts of prescriptive commands would also qualify as types of S-representation.

Yet there is a third sort of rule that doesn't appear to be a special case of either the IO-representation notion or the S-representation notion. This type of rule is thought to encode conditional commands that are couched in purely computational terms, where the rule is not about the simulated target domain but instead about some internal operation that the system itself must perform. Instead of "pick up block" or "move Bishop to such-and-such position," the command is thought to mean something like "perform computational sub-routine Z" or "re-write symbol W in position S." In other words, the command refers to various aspects of the model or simulation process itself rather than to aspects of processes that are being modeled or simulated. Consequently, the content of such a command never goes outside of the realm of the computational system. It is this third notion of rule representation that I want suggest is *not* doing any valuable explanatory work.

While these issues are notoriously tricky, it should first be pointed out that it is actually far from clear that these sorts of computational rules actually *are* part of the explanatory framework offered by the CCTC. Remember that the CCTC is a theory of how cognitive systems work. It is not a theory of how to *implement* the states and processes described by that theory in an actual machine. It could be argued that the structures in computational systems that serve as rules in this sense are really just part of the implementing architecture, and not an essential part of the CCTC's explanatory apparatus.[15] They are perhaps essential for programming actual physical machines, but they aren't essential for understanding the sense in which cognitive processes are said to be computational.

Yet some might say that in certain theories, these types of rules are indeed intended as part of the theory's explanatory apparatus. Let's assume for the sake of argument that this is so. We can see that this "internal" notion of rules cannot play the same sort of explanatory role played by either the IO notion or the S notion. Suppose there is a sub-component of the system with the mechanical job of erasing and writing symbols. Moreover, suppose this sub-component is triggered to erase the symbol "X" and re-write the symbol "Y" by receiving as input yet a third formal token. Do we need to treat this third data structure as representing the command "erase symbol 'X' and re-write symbol 'Y'" in order to treat this sub-component as a symbol eraser/writer? Surely the answer is "no." Because the sub-component is doing purely mechanical operations, we can treat the sub-component's inputs as merely formal tokens and its outputs as actual symbol erasings and writings without treating either the input or output as representations. To view a sub-system as an adder, we need to view its inputs and outputs as representations of numbers. But to view a yet more basic sub-system as an eraser and re-writer of formal symbols, we don't need to treat *its* inputs or outputs as representations of anything. We need to treat its outputs as the erasing of formal symbols, but we don't need to pay any attention to what these symbols represent.

The S-representation notion wouldn't apply to these "rules" either since they don't serve as elements of a model or simulation that the system is using. If anything, the rules are thought to correspond to the mechanical details of the simulation itself – about the simulat*or*, not the simulat*ed*.

[15] See Fodor and Pylyshyn (1988).

Rules of this sort refer (allegedly) to the "behind-the-scenes" mechanical steps or processes that are necessary for the simulation's execution. They aren't themselves *part of* the simulation or model. Hence, this notion of representation can't serve as a form of S-representation either.

If internal rules of this sort (that is, rules that designate specific mechanical operations) can't serve as either IO-representations or as S-representations, then in what sense are they supposed to serve as representations? One proposed answer is that we should treat these internal states as representations of rules simply because they generate various state-transitions in the computational process. Because these structures are causally pertinent, the system is thought to "follow" commands or instructions that they encode. For example, Newell and Simon (1976) offer the following account of what it is for a computational system to "interpret" a symbolic expression: "The system can interpret an expression if the expression designates a process and if, given the expression, the system can carry out the process . . . which is to say, it can evoke and execute its own processes from expressions that designate them" (1976, p. 116). So on this view, executing a process amounts to interpreting a rule or command "expressing" the procedure that needs to be implemented. We view structures as representations of commands because these structures cause the system to carry out the expressed procedure.

The problem with this perspective is that it suggests a notion of representation that is too weak to have any real explanatory value. There is no beneficial level of analysis or explanatory perspective that motivates us to regard things as representations simply because they influence the processing. There is nothing gained by treating them as anything other than causally significant (but non-representational) components of the computational system. Of course, we can always cook up a command corresponding to the relevant causal role, and then allege that the structure represents that command. For example, we can say that a spark plug's firing expresses the rule, "piston, go down now," or that a door-stop represents to the door the command, "stop moving here." But there is no explanatory pay-off in treating these things in this way. Or, putting things another way, spark plugs and door stops don't actually *serve as* representations. Similarly, calling computational elements representations of rules simply because they initiate certain computational operations adds nothing to our understanding of how computational processes are carried out. There is no sense in which states that cause different stages of computational processes actually play a representational role, and nothing is added to our understanding of computational systems by treating *these* sorts of structures as things that encode rules.

The fact that the so-called rules can be modified, so that they have different influences on the processing at different times, is itself sometimes suggested as a justification for treating them as encoding instructions. That is, because the causal influence of a computational element can be altered, it is suggested that this alterability gives rise to their status as representations. But it is hard to see why this should matter. There are plenty of causal systems in which the inner elements can be adjusted but clearly aren't serving a representational role. For example, a commonplace timer mechanism turns lights and other appliances on and off by closing an electrical circuit at specific times. The activation times can be changed by moving pegs to different locations on a 24-hour dial, so the pegs control the timing of the flow of electricity by their position on the dial. Given this modifiable causal role, someone might propose that the pegs in specific slots encode "rules" like, "If it is 6:45 p.m., then turn on the lamp" or "If it is 11:30 p.m., then turn off the lamp." We can, in other words, adopt the intentional stance with regard to the timer pegs, and claim that the timer "interprets" and obeys these commands. However, there is no reason to adopt this perspective. We can understand everything we need to know about how the pegs operate in the timer without introducing representational language. The same goes for causally relevant components of computational systems that are necessary for the implementation of computational processes. Unlike the situation with IO-representations or S-representation, the intentional stance buys us nothing with these structures, and the fact that their influence can be modified doesn't change this. We can understand everything about the way they function in the system – about their purpose and computational significance – without regarding them as rules that are in some sense interpreted by the system.[16]

In chapter 5, we will return to the idea that things serve as representations because of what they cause. For now, the point of this digression is that when

[16] There is yet another feature of computer elements, besides their causal relevance, that invites researchers to regard them as representations of commands. The feature concerns the way these elements are typically created and modified in actual programs, which is by a programmer typing out instructions in a programming language that *we* would translate as something like "when X, do sub-routine Y." It is not so surprising, then, that something is thought to be a representation of such a rule *for the system*, especially since the system appears to be following just such a command. Yet, this is just an artefact of the way computer programs are written, one that doesn't change our earlier verdict that we lack a good reason for positing encoded rules that the system interprets. Suppose we altered the way in which the timer pegs get placed, so that it now happens via typed commands. To get the "on" peg to move to the position that will turn on the lamp at 6:45, I type on some keyboard, "If 6:45 p.m., then turn on lamp." It seems intuitively clear that this modification in the way the pegs get positioned does nothing to provide a justification for viewing them as representations of rules. It might explain why we would be tempted to call them rules, but it doesn't alter the basic fact that their functionality is no different from other basic causal elements.

proponents of the CCTC posit rules that are employed by computational systems, they are often referring to structures that are indeed representations of rules, because they are special cases of either interior IO-representation or S-representation. But sometimes commentators refer to something that is not serving as a representation of a rule at all. In the case of the former, the nature of the CCTC explanations demands we treat these structures as representations of rules; in the case of the latter, it does not.

3.4.4 The vindication of folk psychology revisited

In the last chapter, we saw how the Standard Interpretation links the positing of representations in the CCTC to folk psychology. On the Standard Interpretation computational structures receive their representational gloss by serving as realizers of beliefs and desires. Rather than first demonstrating how the CCTC itself invokes inner representations and then exploring if and how the folk notions map onto this account, this perspective suggests that the CCTC applies the folk notion of representation to computational structures as a way of showing that the notion of representation is needed. Representational states are thereby seen as theoretical add-ons that are not directly motivated by the CCTC explanatory framework. The end result is a picture in which the explanatory value of representation becomes questionable, and the representational nature of the CCTC is called into doubt.[17]

Yet we can now see that on the proper interpretation, representational notions are actually built right into the explanatory pattern offered by the CCTC. IO-representations and S-representations are an indispensable feature of the theoretical framework, and the explanatory value of these notions is independent of anything associated with folk psychology. With this corrected picture of CCTC representation in hand, we can now return to the question of whether or not, if true, the CCTC would provide a vindication of folk psychology.

If we are going to show that a folk concept of some sort is vindicated by a scientific theory, then the first obvious step is to establish that the scientific

[17] Like the S-representation notion, the Standard Interpretation appeals to a sort of isomorphism to establish the need for representations. But it is the wrong sort of isomorphism. The isomorphism it exploits is between the causal structure of symbol manipulations and the sort of psychological processes stipulated by folk psychology. This merely tells us that computational symbols can behave like the posits of folk psychology; it doesn't provide us a reason for thinking those symbols should be treated as representations. S-representation, on the other hand, says the isomorphism that matters is between the symbol manipulations on the one hand, and whatever it is that is being modeled or simulated on the other hand. This sort of isomorphism establishes how computational structures serve as representations because it requires computational structures to serve as components of models and simulations.

theory is actually committed to something with the central features associated with the folk notion in question. Unfortunately, there are no clear criteria for what in general counts as "central features." Nor is there a clear consensus on how many central features need to be possessed by the scientific posit to distinguish cases of retention from cases of elimination (Ramsey, Stich, and Garon 1990; Stich 1996). Consequently, the analysis must be done on a case-by-case basis and unavoidably involves a judgment call. In some cases, reduction only requires that the scientific posit play the same causal roles as the folk posit. But this is often not enough – epileptic seizures don't vindicate demonology even though epileptic seizures cause many of the behaviors associated with demonic possession.

Folk psychology is committed to the existence of mental representations. Therefore, for folk psychology to be vindicated, the correct scientific theory needs to invoke, at the very least, inner cognitive representations as well. What we can now see (but couldn't from the perspective of the Standard Interpretation) is that the CCTC meets this minimal requirement. The CCTC is indeed a representational theory of the mind – one that is committed to the idea that the brain employs structures that are inner representations. If the CCTC is correct, then at least *this* aspect of folk psychology will be vindicated.

Of course, this is only part of the story. The scientific account must also posit representations with the right sort of properties. Since the central properties of propositional attitudes are their intentional and causal properties, the scientific theory must posit representational states with similar intentional and causal properties. If the posits are too dissimilar from our ordinary notions of mental representation, then, despite serving as representations, the psychological theory may be too unlike our commonsense psychology to provide a home for the posits of the latter. For example, Stephen Stich, Joseph Garon, and myself have argued that connectionist distributed representations don't qualify as the right sort of representations because they lack the requisite functional discreteness to act in the manner commonsense psychology assumes of beliefs and propositional memories (Ramsey, Stich, and Garon 1990). Distributed connectionist representations can't vindicate folk mental representations because the former lack the sort of causal properties the latter needs.[18]

[18] I now believe that our eliminativist analysis of connectionist networks didn't go far enough, since my current view is that it was a mistake to allow that distributed networks employ *any* legitimate notions of inner representation. My reasons for this view will be spelled out in the next two chapters. See also Ramsey (1997).

Our current concern is not with connectionism, however, but with the CCTC. Are the notions of IO-representation and S-representation the sort of posits with which beliefs and other folk notions could be identified? While the two computational notions are not the same as the folk notions, they clearly share many of the same features. They both have the sort of intentionality that we associate with our thoughts and they are also capable of the kind of functional discreteness that folk psychology assigns to beliefs and desires. Moreover, in many respects, the sense in which they serve as representations overlaps with the sense in which we arguably think thoughts serve as representations. To see this last point better, consider a piece of folk psychological reasoning that Fodor treats as instructive, offered by Sherlock Holmes in the "The Speckled Band":

> "... it became clear to me that whatever danger threatened an occupant of the room couldn't come either from the window or the door. My attention was speedily drawn, as I have already remarked to you, to this ventilator, and to the bell-rope which hung down to the bed. The discovery that this was a dummy, and that the bed was clamped to the floor, instantly gave rise to the suspicion that the rope was there as a bridge for something passing through the hole, and coming to the bed. The idea of a snake instantly occurred to me ..." (In Fodor 1987, pp. 13–14)

Here Holmes is offering, as Fodor notes, a bit of reconstructive psychology. He is applying commonsense psychology to himself to explain how his realizations, thoughts, observations and ideas led to his conclusion that the victim died of a snakebite. What Fodor asks us to note is how much Holmes's account resembles an argument, with clear premises, conclusions and chains of rational inference. Because classical computational systems are good at this type of formal and explicit reasoning, they provide, according to Fodor, the avenue for vindicating commonsense psychology.

But now consider the same passage from the standpoint of S-representation. Instead of describing a reasoning process that looks like a formal argument, Holmes's account of his own reasoning can be seen as involving something like a model of the events that led to the victim's demise. In this passage, Holmes at least implies that he discovered the solution by mentally reconstructing the critical series of events that were involved in the murder – a reconstruction that included representations of the relevant elements (vent, dummy rope, snake) and the pertinent events (the snake slithering down the rope) to complete the picture and solve the crime. Holmes's version of folk psychology makes it sound a lot like running a simulation of events and processes, or building a model and then, as we say, "connecting the dots."

My point is not to challenge Fodor's Conan Doyle scholarship. Rather, the point is that folk psychological explanations of mental processes can often be seen to characterize those processes as involving models or simulations, or what we earlier referred to as "surrogative reasoning." If this is correct, then folk notions of mental representations may well be very close to the notion of S-representation proposed by the CCTC. The S-representation notion, although not identical to our ordinary notion of propositional attitudes, may well be in the ballpark of the kind of representational state that could vindicate a modified version of folk psychology.[19] While it is hard to see how beliefs could turn out to be mere syntactic states with an unspecified representational role (as suggested by the Standard Interpretation), it *does* seem they could turn out to be representational components of models that our brains use to find our way in the world. Consequently, if the CCTC should prove correct, then that may provide us with good reason to think that belief-like states will find a home in a serious scientific psychology after all. The CCTC may indeed vindicate commonsense psychology, but not without first being understood as a theory that invokes inner representations for its *own* explanatory reasons.

3.5 SUMMARY

It is important to be clear on the objective of this chapter. The aim has not been to defend the CCTC as a true theory of our cognitive processes. Rather, it has been to defend the idea that the CCTC is indeed a *representational* theory of our cognitive processes. My goal has been to show how the CCTC framework makes use of notions of representation that, contrary to the Standard Interpretation, are needed for reasons that are independent of any desire to vindicate folk psychology. As we've seen, one notion is connected to the hierarchically organized, sub-modular nature of cognitive processes generally assumed by the CCTC. The other notion is connected to the sorts of models and simulations many versions of the CTCC paradigm invoke. Both notions of representation appear to meet the job description challenge and reveal how CCTC theories of the mind are

[19] A question some have posed is this: How do folk notions of mental representation meet the job description challenge? Initially, it seems that they clearly don't. Folk psychology doesn't tell us *how* mental states like beliefs come to serve as representations; it simply presupposes their representational status without trying to explain it. This is one of the key differences between folk psychology and many sorts of scientific psychology. Yet on second thoughts, it might turn out that, deep down, our concept of belief includes the role of serving as part of a person's inner model of the world. If this is so, then beliefs would simply be a type of S-representation.

representational theories of the mind. While there are difficulties associated with each of these posits, these are perhaps no worse than the sort of philosophical problems associated with many posits of scientific theories.

This analysis of representational notions that succeed in meeting the job description challenge will serve as a contrast to what comes next. In the next two chapters, we'll look at two different notions of cognitive representation that have become popular among those working in the cognitive neuroscience and connectionist modeling. Unlike my treatment of the notions of representation discussed here, I'll argue that these notions fail to meet the job description challenge and do no real explanatory work. My claim won't be that the non-CCTC theories are false. Rather, my claim will be that, contrary to the way they are advertized, many of these accounts fail to invoke internal states and structures that are playing a representational role. When we look closely at these other representational notions, we can see that the states they describe are not really serving as representations at all.

4
The receptor notion and its problems

In the last chapter we explored two notions of representation associated with the CCTC paradigm and saw how these notions do important explanatory work within this theoretical framework. As we saw, it is possible for CCTC to address the job description challenge and show how notions of representation actually belong in a physicalist (or computationalist) story of how the brain works. By contrast, I'll argue that the notions of representation we are going to explore in the next two chapters do *not* actually meet the job description challenge and do not enhance our understanding of the cognitive systems that allegedly use them. Indeed, I'll suggest that the propensity to regard states as representations for the reasons associated with these notion has led to some deep misconceptions about the nature of many of the newer, non-CCTC theories of cognitive operations. In this chapter, we'll examine one of these notions – or more accurately, a family of notions – that has become a popular way of thinking about mental representation in such fields as cognitive neuroscience, connectionist cognitive modeling, and cognitive ethology. The family of representational notions I'll explore is one I will simply call the "receptor notion." In the neurosciences, the same sort of state is often referred to as a "detector."[1] After explaining what I take this notion to involve, my aim will be to argue that it is not a useful theoretical posit of cognitive science. In effect, I'll argue that things described as representations in this sense are not really representations at all. I will not argue that cognitive systems actually lack states and structures that do the things ascribed to receptor representations. Instead, I'll deny that structures that do those sorts of things are serving as representations.

To show all of this, my discussion in this chapter will have the following organization. First, I'll try to spell out the basic idea behind receptor representation as it appears in different theories of cognitive science. Of

[1] See O'Reilly and Munakata (2000, pp. 24–26).

course, I can't provide an exhaustive survey of the many different theories in which this notion appears. However, as I did in the last chapter, I'll try to provide enough of a sketch to make it fairly clear what people have in mind when they invoke this sort of representational posit as part of their account of cognition. Next, I'll ask how well this notion of representation fares with regard to the job description challenge. As we'll see, the receptor notion is seriously flawed and needs enhancement if we are to make sense of inner states serving as repesentations in this way. To this end, I'll turn to philosophical work on mental representation – in particular, the work of Fred Dretske (1988). I'll try to show that Dretske's own account of representation overlaps a great deal with the receptor notion, yet is significantly more sophisticated and robust. Thus, some have suggested that Dretske shows how the receptor notion can be improved in ways that would allow it to handle the job description challenge. I'll explore two elements of Dretske's account that might be thought to do this – his theory of misrepresentation, and his discussion of how something serves as a representation. I'll then argue that neither one of these aspects of Dretske's account can save the receptor notion and that, in fact, Dretske's own account of representation is equally flawed. After addressing a number of related concerns, I'll conclude by further explaining just why, in light of these considerations, the receptor notion should be abandoned.

4.1 THE RECEPTOR NOTION

At the heart of the receptor notion is the idea that because a given neural or computational structure is regularly and reliably activated by some distal condition, it should be regarded as having the role of representing (indicating, signaling, etc.) that condition. Such structures are viewed as representations *because* of the way they are triggered to go into particular states by other conditions. Unfortunately, in the scientific literature there is not much in the way of explicit exposition of this representational notion, despite its widespread appeal.[2] Consequently, our initial understanding of the notion will need to be gleaned from the way it is employed in the descriptions and discussion of various theories.

One of the most frequently cited sources for the idea of receptor representations is Hubel and Wiesel's (1962, 1968) important work on neurons in the visual system of cats, monkeys and rats. In a series of papers,

[2] Some notable exceptions include Reike *et al.* (1997), O'Reilly and Munakata (2000), and especially de Charms and Zador (2000).

Hubel and Wiesel reported on the excitatory and inhibitory activity of neurons responding to visual stimuli, such as a slit of light presented at different angles of orientation. While they are commonly cited for discovering these so-called "edge detectors," in truth, Hubel and Wiesel are extremely cautious about ascribing any sort of representational role to these neurons, and primarily restrict their discussion to their response profiles and receptive fields. By contrast, in their famous paper, "What the frog's eye tells the frog's brain," Lettvin, Maturana, McCulloch, and Pitts (1959), quite explicitly endorse the idea that certain neurons should be viewed as "detectors" precisely because they reliably respond to certain stimuli. Toward the end of their article, they note,

> What, then, does a particular fiber in the optic nerve measure? We have considered it to be how much there is in a stimulus of that quality which excites the fiber maximally, naming that quality . . . We have been tempted, for example, to call the convexity detectors "bug detectors." Such a fiber responds best when a dark object, smaller than a receptive field, enters that field, stops, and moves about intermittently thereafter. (1959, p. 1951)

This basic idea, that neural structures should be viewed as having the job of detecting[3] or representing because they respond in a reliable manner to certain conditions, has become a common assumption in the cognitive neurosciences. For example, in an article on the role of neurons in perceptual processing we are told, "[I]f a cell is claimed to represent a face, then it is necessary to show that it fires at a certain rate nearly every time a face is present and only very rarely reaches that rate at other times" (Barlow 1995, p. 420). While this author refers to the necessary conditions for claiming a cell serves to represent faces, it is clear from the literature that this sort correlative response profile is regarded as a *sufficient* condition as well. Indeed, researchers often skip the question of *whether* neural receptors function as representations, and instead ask about how the actual representational encoding is done. That is, researchers often begin with the assumption that neurons function as representations, and then explore, for example, whether the encoding is in single-cell "grandmother" representations, or instead distributed across a population of neurons. In more recent

[3] For some, there might be subtle differences, at least in connotation, between "detector" and "representation." Here I'm treating detection as a type of representation because that is actually how most cognitive investigators view them. As O'Reilly and Munakata note, "the neuron that detects an oriented bar at a given position is said to *represent* an oriented bar at a given position" (2000, p. 25).

neuroscientific theories, this idea has been supplemented by introducing more formal theories of information, such as the one suggested by Shannon and Weaver (1949). Since neurons often stand in some sort of nomic dependency relative to specific stimuli, they are thereby thought to qualify as "information carriers." Throughout many areas of the neurosciences today, especially in neuro-biological accounts of perception, representation or information carrying are often treated as a basic task that neurons perform.

A similar notion of representation is increasingly becoming commonplace in many of the newer forms of artificial intelligence. In particular, in connectionist modeling, the internal "hidden" units of networks are generally characterized as playing a representational role because of the way they respond to input to the network. A common task of multi-layer networks is to learn to discriminate between varying forms of stimuli. As a result of this learning, specific activation patterns of the internal units become correlated with specific types of input. If sonar echoes from undersea mines tend to generate patterns of one sort, whereas echoes from rocks generate patterns of a different sort, then those distinctive patterns are thought to function as internal "distributed representations" of mines and rocks (Gorman and Sejnowski 1988). Connectionist modelers often note the similarities of these internal units to neural receptors, suggesting they provide a more biologically plausible form of representation than traditional computational symbols.

In fact, this perspective is often bolstered by mathematical analyses of networks that are thought to reveal the concealed ways in which these internal units actually serve as representations. We can treat the activation level of each internal unit as an axis in a multi-dimensional hyperspace, so that any given overall pattern of activity corresponds with a point in that space. These points are then correlated with the input that produced them and we can get a sense of how the network is responding to different types of stimuli. The results of these analyses have generated considerable speculation and are thought to illustrate the "conceptual" scheme discovered by the network. For example, in NETtalk, a connectionist model trained to convert representations of written text to phonemes, the patterns generated by written vowels were clustered in one main group while those generated by consonants were collected in a different region of vector space (Sejnowski and Rosenberg 1987). Each major grouping also consisted of smaller clusters revealing that hidden unit activation patterns triggered by a "p" were next to those triggered by a "b", but some distance from those triggered by a "g", as illustrated in figure 4a.

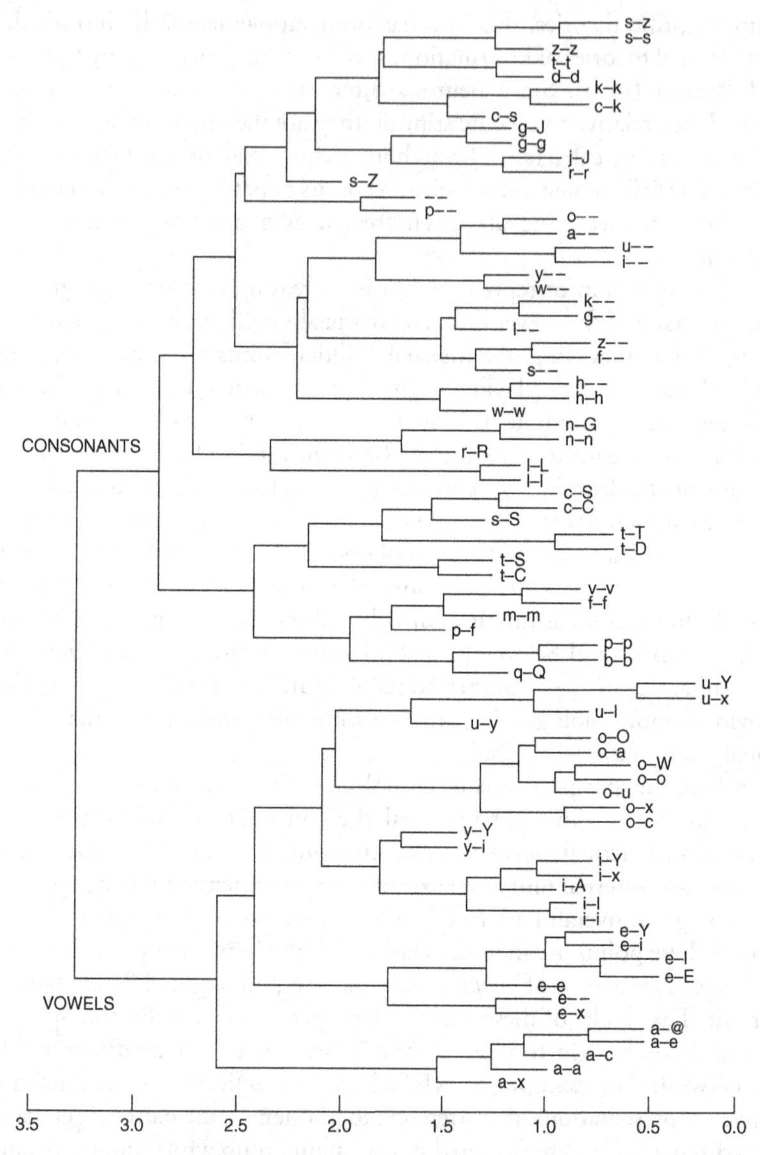

Figure 4a Multi-dimension state-space representation of the response profile of NETtalk's hidden units. The position of letters on the graph corresponds to the position in vector space of the hidden unit activation pattern generated by that phoneme. The closer the letter positioning, the more similar the corresponding activation patterns. Thus, the patterns generated by "f" and "v" are quite similar, whereas the pattern generated by "o" and "t" are quite dissimilar. From Sejnowski and Rosenberg, 1987. Copyright owned by Wolfram Research Inc., reprinted with their permission.

The organizational structure of these groupings, and the fact that they seem to cluster around a prototype, have prompted many writers to conclude that the hidden units perform a representational function. Thus, Churchland and Sejnowski tell us that, "all the vectors for vowel sounds clustered together, indicating that they were represented in the network by patterns of activity in units that were distinct from those representing the consonants (which were themselves clustered together)" (1989, p. 35).

Another area where the receptor notion of representation appears quite frequently is cognitive ethology and studies of simple organisms. We have already seen how Lettvin *et al.* have characterized neurons in the frog's brain as representations of bugs because of the way they respond to bug-type displays. The receptor notion also appears in accounts of the cognitive processes and perceptual systems in more sophisticated animals. For example, arguing for the existence of "face cells" in monkeys, Elliffe presents neurophysiological studies that "have identified individual neurons in the inferior temporal region of the macaque brain which fire in response to visual presentation of faces . . . Ensembles of such cells can thus be said to code for the presence of individual face stimuli or, less formally, to 'recognize' them" (Elliffe 1999). Yet not all cases of receptor representation in non-humans involve neuronal states. In one famous example, researchers have suggested that iron traces employed in the propulsion mechanism of certain anaerobic bacteria serve as magnetosomes. Because these iron particles are reliably pulled in the direction of magnetic North and, as it happens, toward oxygen-free water, they are viewed as tiny compass-like representations that actually tell the bacteria which way to go (Blakemore and Frankel 1981).

There are, of course, many more examples of the receptor notion employed throughout the cognitive sciences, but this should suffice to get across the basic idea. Some receptor representations are assumed to exist deep within a given cognitive system, whereas others are more on the periphery. Some involve the responsiveness of individual structures while others involve ensembles of neurons reacting to stimuli together. What we see with all of these is a common fundamental conviction that, roughly, an important type of cognitive representation results when some sort of internal state reliably responds to, is caused by, or in some way nomically depends upon some external condition. When this occurs, such a state, whether it be called a "representation" or a "detector" or an "indicator" or some such, is viewed as having the role of representing that external condition because of this causal or nomic dependency relation.

To philosophers with some awareness about naturalistic theories of meaning, this should all sound quite familiar. The central idea behind the receptor notion is an old one that has been discussed in philosophical circles for some time. In chapter 1 we discussed Pierce's notion of "indices," a form of representation based upon some sort of dependency relation between the sign and its object. The same basic principle is at work in the receptor notion, though the receptor notion, unlike Pierce's signs, is thought to serve as representation without the inclusion of the third-party interpreter. More recently, philosophers have developed the idea that there is a sort of "natural meaning" or "informational content" that results from the way a state reliably co-varies with some other state of affairs. It is this notion of representational content, made famous by Grice (1957), that we exploit when we use tree rings to inform us of the age of a tree, or that prompts us to say things like, "smoke *means* fire." Moreover, several philosophers have developed their own theories of content for mental representations by building upon this basic idea. Below, I will say more about the way one of these philosophical projects links up with the receptor notion. But first, I want to evaluate the receptor notion in terms of the job description challenge and demonstrate that, as it stands, the notion is severely deficient in answering this challenge.

4.2 THE RECEPTOR NOTION AND THE JOB DESCRIPTION CHALLENGE

Recall that the job description challenge involves a specific condition that needs to be met if a theoretical notion of representation is going to be explanatorily useful. Besides some sort of account of what determines the content for a given state, we also (and perhaps more importantly) need an account of how the structure or state in question actually serves as a representation in (and for) the system. So for example, in the case of S-representation, the state serves as a representation by serving as an element of a model or simulation of the target domain in a computational system using surrogative reasoning. What we need is a similar sort of account for receptor representation that manages to give us a sense of how receptors play a role that is recognizably representational in nature.

Given this, we can see that the receptor notion faces a *prima facie* difficulty. The problem is that the receptor notion does not appear to provide us with an account that reveals why a given state or structure should be seen as serving as a representation. That is, while the receptor notion provides us with what could perhaps serve as an account of a content-determination relation (nomic dependency), it does not provide us with any sense of how a state or structure actually plays a representational role. Their actual stated functional role – reliably responding to some external condition – is not, by itself, a role sufficiently representational in nature to justify treating some state or structure as a representation. There are several *non-representational* internal states that must, in their proper functioning, reliably respond to various states of the world. Our immune system, to take one example, functions in part by consistently reacting to infections and insults to our bodies. Yet no one suggests that any given immune system response (such as the production of antibodies) has the functional role of representing these infections. While nomic dependency may be an important *element of* a workable concept of representation, it clearly is not, by itself, *sufficient* to warrant viewing an internal state as a representation. Something more is needed if we want to invoke a notion that is useful and that belongs in a theory of cognition.[4]

This complaint against the receptor notion can be understood as involving the following dilemma. On the one hand, it may be claimed that to serve as a sort of a representation *just is* to serve as a state that reliably responds to specific stimuli. But this analysis leads to all of the problems discussed in chapter 1 with overly reduced accounts of representation. We would have the problem of pan-representationalism, because lots and lots of things (e.g., immune system responses) could be said to function as representations in this sense. Moreover, the notion would be reduced to something uninteresting and utterly disconnected from our ordinary understanding of what a representation is. The initially substantive idea that cognition is a process that involves representational states would now become the remarkably boring thesis that cognition is a process that involves states that are triggered by specific conditions. On the other hand, it might be claimed that the conditions in virtue of which receptors serve as representations include factors that go substantially beyond the mere fact that they reliably respond to specific stimuli. But then it is incumbent upon those who invoke this notion to explain what those factors are. Without being informed of these factors or conditions, a critical aspect of representation would be left unexplained and we would have no

[4] Here's how van Gelder puts it: "One of the few points of general agreement in the philosophy of cognitive science is that mere correlation does not make something a representation" (van Gelder 1995, p. 352). Actually, it is far from clear that this is a point of general agreement, as we will see.

clear sense of how states reliably responding to certain stimuli are supposed to function as representations.

Another way to see the problem is to contrast the receptor notion with the notions examined in the last chapter, such as the S-representation notion. What S-representation has going for it that receptor representation lacks is a distinctive role within a cognitive system that is recognizably representational in nature and where the state's content is relevant to that role. It is hard (if not impossible) to see how there could be an account of something serving as a representation without a corresponding reason for treating the content – or, if you like, the fact that the representation *has* intentional content – as an explanatorily relevant fact about that state. With S-representation, the fact that a given state stands for something else explains how it functions as part of a model or simulation, which in turn explains how the system performs a given cognitive task. By contrast, the receptor notion typically carries no similar motivation for treating a state as doing something intuitively representational in nature. There is nothing about way the states are triggered by the environment, or the embedding cognitive architecture, or the explanatory strategy employed, that drives us to treat these structures as representations. Apart from the misleading terminology, their actual functional role suggests we should view them as reliable causal mediators or perhaps relay switches, triggered by certain environmental conditions. They have the function of causing something to happen in certain conditions. But then, many things function this way while properly performing their duties. This alone gives us no reason to treat them as representations.

It should be noted that other authors have, in different contexts, discussed concerns closely related to the ones I'm raising here. For example, Fodor discusses the problem of "pansemanticism" as a worry that arises from the idea "that meaning is just *everywhere*" and "is a natural conclusion to draw from informational analyses of content" (1990, p. 92). And van Gelder (1995) has also directly challenged the idea that mere causal relays are serving a representational role – a challenge we will look at more closely in chapter 6. But despite their awareness of this issue, philosophers have generally not framed their concern in terms of representation *function*. In fact, much of the philosophical discussion does not dwell on what it means for a state or structure to serve as a representation, focusing instead on whether an account of content can be provided that is sufficiently naturalistic. I suspect one reason for this is a tendency to assume that an account of content *just is* an account of representation. However, as we've seen throughout our analysis, a theory of content is only one element of what is needed. We also need an account of *how*, exactly, something plays a representing role in a causal/physical system. In the philosophical literature, there are not many accounts that tackle this issue head-on. One important exception is the ingenious account of representation offered by Fred Dretske.

4.3 DRETSKE TO THE RESCUE?

In the last two sections, I've spelled out the core idea behind the receptor notion of representation as it appears in many cognitive theories, especially many of the non-classical accounts. I've also suggested that to meet the job description challenge, the notion would need to be significantly enhanced to warrant the view that functioning as a receptor entails functioning as a representation. To see how this enhancement might go, the best place to look would be various philosophical accounts of representation that involve the receptor notion. After all, the scientific theories that employ the receptor notion were never intended to provide a robust philosophical defense of their explanatory value. If we are going to give the receptor notion a fair hearing, we need to look at philosophical accounts that are specifically tailored to handle concerns like those raised by the job description challenge.

Unfortunately, finding a philosophical defense of the receptor notion of representation, *per se*, is not so simple. Unlike the case with S-representation, there has been much less in the way of careful, detailed discussion directed at defending this notion as a theoretical posit. The problem is not that philosophers have ignored the ideas associated with this notion; rather, the difficulty stems from the focus upon content noted above. Most philosophical analyses are devoted to providing a content-determining relation for internal states that is sufficiently naturalistic. In this regard, co-variation or nomic dependency relations have taken center stage – not for the defense of a particular type of representational notion, but rather to explain how something presumed to be functioning as a representation acquires its content. Consequently, while the core features of the receptor notion have found their way into philosophical accounts of mental representation, it is far from obvious that many writers would describe their task as one of defending receptor representation, as such.

However, having said all that, I think there are some philosophical projects that come very close to providing what we are after. That is, some philosophical projects ostensibly devoted to providing a naturalistic account of mental representation content, in fact, provide what looks a great deal like a philosophical defense of the receptor notion of representation. Of these, perhaps the most notable and promising is the work of Dretske (1988). Dretske offers an ambitious account of mental representation that is designed not only to provide a naturalistic account of content, but also to show how content is relevant to the production of behavior. While Dretske's stated focus is upon beliefs (or as he puts it, "proto-beliefs") I believe his project can be easily viewed as an account of low-level representational states that are part of the family I have been referring to as receptor representations. Indeed, his theory is clearly motivated by examples of the receptor notion, and many have appealed to Dretske as a way of defending receptor-style representations. What's more, because Dretske's account of content is so closely intertwined with an account of what it is for something to *function as* a representation, we see that he is, indeed, worried about providing a solution to what I have been calling the job description challenge. Thus, if anybody has given a carefully worked-out philosophical explication and defense of the receptor notion – one that tackles the issue of how receptors serve as representations head-on – it is Drestke. Consequently, showing what is wrong with Dretske's account will help us see the fundamental problem with receptor representation in general.

At the heart of Dretske's analysis of mental representation is the notion of "indication," a relation based upon a law-like dependency in which the state of the indicator depends upon the state of that which is indicated. For condition C to indicate another condition F, C must stand in a relation to F characterized by subjunctives like the following: If F had not occurred, C would not have occurred. For example, tracks in the snow indicate a pheasant has passed because if a pheasant had not passed through, there would be no such tracks. This is roughly the same notion of informational content that, as we've seen, forms the basis of the receptor notion.[5] Thus, Dretske's account is built upon the same sort of content-grounding relation that is at the center of receptor representations.

[5] It should be noted that Dretske's account apparently differs in one respect; namely, for Dretske, the dependency or reliability must be 100 percent. If C is triggered by F, say, only 80 percent of the time, then C is not an indicator of F. However, this difference will be largely irrelevant for our discussion.

Yet as we saw in the last section, mere nomic dependency is insufficient to bestow full-blown representational status on cognitive structures. Dretske recognizes this, but for him the inadequacy is not initially couched in terms of whether or not something is serving as a representation. Instead, Dretske focuses on the problem of misrepresentation and the unique form this problem takes in the case of receptor-style representations. The problem – sometimes called the "disjunction problem" (Fodor 1987) – concerns the difficulty in accounting for misrepresentation when a state's content is based upon the way it is triggered by distal conditions. If we assume that the critical content-bearing link between a representation and its intentional object is a dependency based on some form of causal connection, then it is difficult to see how there could ever be such a thing as misrepresentation. We want to say that when a frog's inner fly receptor representation is triggered by a flying BB, then we have a case of misrepresentation. Yet if the inner state is caused by both flies and flying BBs, it seems we really *shouldn't* say that it represents the presence of flies, and misrepresents when triggered by BBs. Instead, we should say that it accurately represents the presence of flies *or* BBs. Hence, the purely causal or nomic dependency story does not seem to allow for falsehood and, thus, fails to provide an adequate account of representational content. Finding a solution to this problem is often treated as the key to naturalizing content. As one author puts it, "the core of any theory of representation must contain an explanation of how misrepresentation can occur" (Millikan 1995, p. 186).

To handle this problem, Dretske joins others – such as Millikan (1984) and Papineau (1984) – and introduces a teleological component to his account, thereby placing a tighter restriction on the sort of causal relations that matter. According to Dretske, internal indicators are elevated to full-blown mental representations when they are recruited – either through natural selection or some sort of learning – as a cause of certain motor output *because of* their relevant nomic dependencies. In other words, indicators become representations when they are incorporated into the processing so that they have a *function* of responding to certain conditions. Their job is to ensure that certain bodily movements are produced whenever the indicated condition obtains. So, in the example discussed earlier, the magnetosomes in anaerobic bacteria indicate the direction of magnetic north, which also happens to correlate with deeper, anaerobic water. Through a process of natural selection, these magnetosomes come to be wired to the bacteria's navigational system because of their nomic link to anaerobic water. They are thus given the functional role of indicating the direction of anaerobic water and, according to Dretske, thereby become anaerobic water representations.

This appeal to functional role does two things for Dretske. First, it enables him to handle the problem of misrepresentation in an intuitively appealing way. Error is now possible because we can say that whenever an indicator is triggered by something other than what it is *supposed to* indicate – something other than what it was, in fact, selected to indicate – it is a case of misrepresentation. If the magnetosomes react to something other than magnetic north (and thus anaerobic water), we can say they are misrepresenting since they are responding to something other than what they were recruited to indicate. Second, it provides Dretske with a way of showing how informational content can be explanatorily relevant. Structures are recruited as causes of motor output *because* they indicate certain conditions. Thus, being an indicator is a causally relevant feature of a structure, and since Dretske regards the indication relation as a type of natural meaning, he suggests a type of *meaning* is causally relevant. Informational content serves as what Dretske calls a "structuring cause" – a feature that helps explain why a structure becomes incorporated into a system's functional architecture. It is in this way that Dretske believes meaning is explanatorily relevant and justifies our viewing a structure as having a representational function.

So on Dretske's account, to serve as a representation something must a) stand in some sort of nomic dependency relation to some distal state of affairs, and b) become incorporated into the processing *because* of this dependency, thereby acquiring the function of indicating those states of affairs. While there are a number of further details and elements to Dretske's theory, for our purposes, the relevant question is whether there is anything here that might serve as the "something more" that could be used to beef up the receptor notion of representation and answer the job description challenge. Many philosophers and cognitive scientists familiar with Dretske's work have assumed the answer to this question, or a least a similar one, is "yes" (see, for example, Bechtel 1998). Indeed, there appear to be at least two components of Dretske's account that could be thought to successfully enhance the receptor notion. The first is Dretske's teleological strategy for modifying natural meaning in order to solve the misrepresentation problem. The second is Dretske's account of how content comes to play an essential role in the explanation of behavior. What I would like to do next is take a very close look at each of these elements. I plan to show that despite their initial appeal, neither of these factors can serve as an adequate answer to the job description challenge for receptor representations. Consequently, the factors that Dretske provides do not help to elevate the receptor notion to a legitimate, explanatorily valuable notion of cognitive representation.

4.3.1 Does Dretske's solution to the disjunction problem help solve the job description challenge?

We saw above just how important the issue of misrepresentation is for a working theory of content. Because the capacity for misrepresentation is considered a hallmark of intentionality, it is often assumed that the key to naturalizing content (and providing a theory of representation) is to explain in causal or physical terms how a state can succumb to error. Dretske himself notes, "it is the power to misrepresent, the capacity to get things wrong, to say things that are not true, that helps *define* the relation of interest. *That* is why it is important to stress a system's capacity for misrepresentation. For only if a system has this capacity does it have, in its power to get things right, something approximating *meaning*" (1988, p. 65). Hence, for many philosophers, explaining how something serves as a full-blown representation has largely been a matter of showing how something can misrepresent. But does an account of error provide us with sufficient conditions to regard receptors not just as causal mediators, but as full-blown representations?

In answering this question, we need to bear in mind the specific way in which Dretske and others appeal to teleology to solve the problem of misrepresentation. The relevant notion that does all the work is the notion of something *malfunctioning* – of some device or element not playing the functional role it was designed or selected to perform. Of course, *by itself* the notion of malfunction (like the notion of function) has no intentional aspect or implication. Errors of this sort are ubiquitous throughout biological systems – all sorts of states with non-representational functions are capable of malfunctioning. Indeed, all sorts of states with non-representational functions are capable of malfunctioning by being triggered by inappropriate or aberrant causes. So we don't get *semantic* notions of error, like falsehood or misrepresentation, through an appeal to teleology alone. One must *first* assume that the state in question is serving some sort of representational function, and *then* we can appeal to teleology to tell a story about how error arises through malfunction.

Consequently, the teleological solution to the misrepresentation problem cannot count as a solution to the job description challenge because the former presupposes that the latter has already been given. That is, to solve the misrepresentation problem in the manner suggested by Dretske and other teleo-semanticists, we must *begin* with the assumption that the state in question serves as a representation. Then (and only then) we can appeal to notions of malfunctioning to account for misrepresentation. The job description challenge raises the following question: "By virtue of what, exactly, is a state serving as a representation as opposed to something else?" However, the disjunction problem raises a different question, namely, "How does a representational state come to have a specific, appropriate sort of content as opposed to some sort of aberrant, disjunctive content; how do we show a neural state represents flies and not flies or BBs?" Dretske provides a reasonable answer to the second challenge; he limits *what* a state represents by appealing to what it *ought* to represent. But this leaves untouched the first question about whether the state in question represents *at all.*

4.3.2 Does Dretske's account of representation function help?

If Dretske's account of misrepresentation cannot provide the "something more" needed to shore up receptor representation, then what about his account of how representations acquire their functional role? After all, a story about how a state comes to function as a representation is the very thing we are after. Moreover, it is through this process that, according to Dretske, the structure's informational content comes to play an important explanatory role. So perhaps it is here that Dretske gives us something that can supplement the receptor notion and provide an answer to the job description challenge.

Recall that for Dretske something becomes a full-blown representation when it acquires the function to indicate, and the key to acquiring that function is being recruited as a cause of some sort of physiological response because of the indication relation. In other words, a structure becomes a full-blown representation when it is recruited as a cause of motor output because it reliably responds to some sort of condition. As Dretske puts it,

> These internal indicators are assigned a job to do in the production of bodily movement . . . in virtue of what they "say" (indicate or mean) about the conditions in which these movements have beneficial or desirable consequences . . . Learning of this sort mobilizes information-carrying structures for control duties *in virtue of the information they carry.* (1988, pp. 98–99)

Dretske clearly assumes that if the relevant nomic dependencies are relevant to the proper functioning of the structure, then whatever information is carried by those dependencies must be relevant as well. This assumption – that the explanatory relevance of information or natural meaning is apparently *entailed by* the explanatory relevance of some sort of nomic dependency – is also a critical element of the receptor notion of representation. It is perhaps the main motivation behind the common tendency to treat things that *respond* as things that *represent.* If neural structures are actually recruited as causes of bug-catching movements because they are reliably caused to fire by the presence of bugs, then it certainly seems tempting to assume that they are serving as bug representations. The question we need to address is, does this arrangement suffice for something to serve as a representation?

To answer this question, it will help to try to get a clearer understanding of the relation between the purely causal/physical or nomic dependencies that are thought to "underlie" the indication relation, on the one hand, and the quasi-semantic, informational relation often said to be "carried by" these dependencies on the other hand. Dretske and many authors are somewhat unclear on the nature of this relation. While it is fairly clear what it means to say that state A nomically depends upon state B, it is much less clear how such a claim is supposed to translate into the claim that A is an indicator of B, or how we are to understand expressions like "information flow" and "information carrying." If A indicates or carries information about B, is this property supposed to be in some way distinct from A's being nomically dependent upon B, perhaps supervening on or emerging from that nomic dependency? Or is it instead the case that when we say A indicates or carries information about B, this is to be translated as identical to the claim that A is nomically dependent upon B; that is, are these just two different ways of referring to the same condition?

On the former view, the information carried by A can be understood as something that is somehow separate and distinct from the other relational properties of A. If A carries information about B, this is a distinct property of A, whether or not anyone or anything exploits it. On this analysis, the term "indication" would be ambiguous – perhaps sometimes referring to the non-semantic nomic dependency, and sometimes referring to something more semantically charged, like information. There would actually be three separate components to Dretske's notion of representation: 1) the non-semantic nomic dependency or causal dependency between A and B, 2) the information about B that results from this dependency and in some sense is carried by A, and 3) the use or exploitation of this information as the indicator is incorporated into the processing. Thus, information is part of the ontology of the world, much like other emergent properties, such as the property of being alive. We can thus call this the "realist" interpretation of information and indication.

According to the second interpretation, terms like "indication" and "indicator" are unambiguous and just mean something like, "nomic dependency" and "reliable responder," respectively. All talk about information carrying would be understood as a way of saying that law-like relations between states of affairs have the potential to be used to discover certain facts. When we say that A carries information about B, we do so because certain states of A nomically depend upon states of B, and thus we (or someone or something) can use A to make warranted inferences about B. That is, talk of information should be interpreted as just a short-hand way of saying that when one state reliably depends upon some other state of affairs, the former can be used to find out things about the latter. On this interpretation, there are only two components to Dretske's account: (1) the nomic dependency A has upon B, and (2) the particular use of A and this dependency to learn something of, or make inferences about, the nature of B. There is, strictly speaking, no further thing that is "carried" by A or "flows" from B to A. Of course, if we so choose, we can certainly call states or structures that are used in this way "information carriers." However, we shouldn't allow this way of talking to mislead us into thinking that there is something over and above the nomic dependency relation and this particular way it is used. There exist various entailment relations between facts that stem from physical conditions of the world, and there is the very special use of this set-up resulting in the acquisition of new comprehension. There is nothing to being an "information-carrier" beyond this arrangement. We can call this the "deflationary" understanding of indication or information.

On separate occasions, I have proposed each of these possible interpretations. On each occasion, I was assured that I was misunderstanding matters and that the alternative reading was correct. Hence, I propose to explore both ways of understanding the indication relation to see if either one can rescue the receptor notion. As we'll see, whichever analysis we adopt won't really matter; in the end the same basic problem plagues both analyses. The central problem is this: Dretske's account of representation appears to assume that if a given structure is incorporated into a system's processing because it nomically depends on a certain state of affairs, it automatically follows that it is being used to stand for (or relay information about) that state of affairs. However, this assumption is unsupported and, as we'll see, is almost certainly false. A structure can be incorporated into the cognitive architecture because it is reliably dependent upon some external condition and yet be doing something besides representing. To see this, I'll first develop the criticism for the realist interpretation of information, and then reformulate it to apply to the deflationary account.

On the realist interpretation, carrying information about a certain state of affairs is in some way different from (though clearly dependent upon) the property of having states that nomically depend upon the same state of affairs. What we need to find out is whether the latter can be explanatorily relevant without the former being explanatorily relevant. That is, can there be situations in which the nomic dependency alone is the reason a given structure is incorporated into the cognitive processing, while the information relation remains irrelevant as a structuring cause?

While Dretske himself never explicitly addresses this question, if we interpret him as adopting the realist position, the answer he seems to presuppose throughout his writings is an unequivocal "no." As we just noted, Dretske clearly assumes that if the relevant nomic dependencies (that carry information) are relevant to the proper functioning of the device, then the information they carry must be relevant as well. For example, in arguing for the causal role of meaning, Dretske argues that information is causally relevant (as a structuring cause) by showing that nomic dependency relations are causally relevant. The triggering mechanism of the Venus fly-trap "signals" the presence of insects because its states are determined by certain kinds of movements. Consequently, Dretkse tells us, "there is every reason to think that this internal trigger was selected for its job *because* of what it indicated, because it 'told' the plant what it needed to know . . . in order to effectively capture prey" (1988, p. 90).

However, it is far from clear why we should think the internal trigger has the job of "telling" the plant anything. While the nomic dependency may serve as a structuring cause in such cases, it doesn't follow that any sort of information or natural meaning is a structuring cause. If these two properties (being nomically dependent and carrying information) are in fact distinct, as the realist interpretation maintains, then there is no reason to think that structures recruited because their states have the property of being nomically dependent on some condition are also recruited because they carry information about that condition. At the very least, we need a further argument to establish that whenever a cognitive system recruits a reliable causal mediator to fill a gap between distal conditions and motor output, the mediator is automatically functioning as a representation. After all, a mediator could be just that – a causal mediator that serves to bring about a desired result in certain specific conditions.

To see how a nomic dependency relation can be important for something's function without the associated information being important, consider the following example. Suppose I decide that I want shade on my back porch at a certain time in the afternoon. Paying close attention to the angles of the sun, I decide to plant a tall tree in a new location, knowing that its shadow will fall exactly where I want, when I want. Here, I would be exploiting a certain nomic regularity – between the length of the tree's shadow and the position of the sun – to assign the tree (or its shadow) a certain job; namely, to keep me cool. On the realist reading of information, the tree's shadow also carries a considerable amount of information. The tree's shadow can be used to learn all sorts of things about the position of the sun and, with enough background information, I could use this arrangement as a crude sundial to tell me the time of day. However, given the way I am *in fact* using the tree's shadow, the information it carries is entirely irrelevant to its functional role. While the nomic relation that exists between the position of the sun and length of the shadow is relevant to explaining why the tree was planted and how the tree is employed, the fact that this nomic relation gives rise to information is not.

Bearing these considerations in mind, we can now see a fundamental problem with Dretske's own cases of so-called representational systems. Consider the well-known example of a bi-metallic strip in a thermostat. If we look closely at the functional architecture of a thermostat, treated by Dretske as an example of an artificial representational system, it is far from clear why we should say that information carrying is the functional role of the strip, as opposed to reliably responding to certain conditions. Inside the thermostat, the strip is rigged up so that it switches on the furnace whenever the ambient temperature drops to a specified level. Thus, it serves as a reliable causal mediator between low temperatures and furnace activation. The functionality of the strip is to cause something to happen in very specific conditions. But in this respect, it is no different than many other devices that we ordinarily treat as having *no* representational function. The firing pin in a gun similarly bridges a causal gap between the pulling of the trigger and the discharge of the round. It also serves to reliably mediate between two distinct states of affairs – to reliably go into a specific state when and only when a certain condition obtains. However, no one thinks the firing pin serves as some sort of representational device. On the realist interpretation, we can indeed say that both the bi-metallic strip and the firing pin carry information. But the information carried by the strip (like the information carried by the firing pin) is not relevant to the job it performs inside a thermostat. Of course, the strip *could* play such a role. If we used the position of the strip to tell *us* about the ambient temperature, then its informational content would be vital to that role. But as it functions in a thermostat, as a condition-dependent trigger for furnace activation, there is no more reason to think that the strip serves to inform the furnace of the temperature than there is to think the firing pin serves to inform the shell about the status of the trigger.[6]

Indeed, it appears the same thing can be said about nearly all of Dretske's cases of alleged representational systems. The triggering devices of carnivorous plants, the magnetosomes in anaerobic bacteria, the so called "bug detectors" of a frog's brain – none of these receptor-type examples provide any principled motivation for claiming that the structure in question is functioning to *represent, transmit information about, stand for*, or *stand in for* something else. For instance, the iron deposits that serve as magnetosomes in anaerobic bacteria are wired to the bacteria's propulsion devices because of the way they reliably respond to anaerobic conditions. They thereby serve as go-betweens or interagents, forging a causal connection between anaerobic water and propulsion devices. We need some further reason, however, for thinking they are recruited into service because of the *information* that results from this relation. There is really no sense in which the bacteria's flagella (their propellers) exploit the informational content carried by the magnetosomes – no sense in which they use the magnetosomes to *stand for* something like anaerobic conditions. It is one thing to serve as a causal mediator between A (anerobic conditions) and B (directional propulsion), it is an entirely different thing to represent A in some way for the system.

It is important to be clear on what the problem is with Dretske's account. The problem is *not* that a structure or device can be employed in such a way that the nomic dependency itself – that enables it to carry information – is explanatorily irrelevant. For example, I'm not claiming that the bi-metallic strip found in a thermostat could also be used as, say, a Christmas tree ornament. Instead, the problem is that a structure can be employed in such a way that the causal and nomic relations that enable it to

[6] In personal communication, Dretske has protested that there is an important difference between the bi-metallic strip on the one hand, and mechanisms like the firing pin on the other hand. In the case of the strip, the device is recruited for the job specifically because of a pre-existing nomic dependency relation (between it and changes of temperature). With devices like firing pins, on the other hand, the nomic dependency results only after it is incorporated into the functional architecture of the mechanism. Yet it is hard to see why this difference should really matter. First, in most realistic biological examples, the relevant nomic dependency would also result from (and not occur prior to) the recruitment process. Moreover, even if structures like firing pins stood in some sort of natural co-variation relation to things like triggers, and were recruited for this reason, it doesn't seem that this should modify their job description. The point is that it is the nomic dependency alone (not information carrying) that matters for the devices' functional role; whether that dependency exists before or after the structuring process seems irrelevant to that point.

carry information *are* explanatorily relevant while the information resulting from such relations is not. In other words, a structure can be employed *qua* nomic-dependent or *qua* reliable-respondent without being employed *qua* information-carrier or, more to the point, *qua* representation. In fact, I'm claiming that this is what happens in the *normal* operation of many of the things treated by Dretske as representations.

So, on the realist interpretation of information, we can say that what Dretske (and others) have failed to explain is how a receptor-type representation is serving as a representation as opposed to serving as something like a reliable relay switch. The mistake lies in thinking that because a nomic dependency relation is a structuring cause, the resulting information is also a structuring cause. But what about the deflationary account of the information relation? My criticism of the realist position rested on the idea that information can be treated as distinct from the nomic dependency relation, and thus the latter can explain the function of a structure without the former explaining anything. But with the deflationary account, there is no such independent informational relation. The indication relation just is the nomic dependency relation. How would our criticism go if we assumed the deflationary stance toward informational content?

On the deflationary reading, the matter depends upon the way we view the functional relevance of a structure's nomic dependency. Recall that on the deflationary view, talk of information carrying is simply talk about a structure's nomic dependency being used in a certain way; namely, as a vehicle for relaying certain facts or making certain inferences. On this reading, the assumption made by Dretske and many cognitive theorists is that if a structure is incorporated into the processing because it (or its states) nomically depend on proximal conditions, we are thereby warranted in assuming that the structure is serving as an informer about those proximal conditions. The relevant question now is whether or not the same nomic dependency can be used for other purposes; if so, then this (tacit) assumption would be mistaken.

As we just saw, it is not at all difficult to show that a structure can realize a functional role that is based upon its states being nomically dependent upon certain conditions, and yet, at the same time, is not playing a functional role that is recognizably representational in nature. The tree and its shadow are recruited into my backyard because of a reliable connection they have to sunlight at certain times of the day. The way the shadow's length is determined by the position of the sun matters a great deal to the job I want it to do. But the job I want it to do is to keep me in the shade, not to tell me anything. The firing pin in a gun functions as a reliable responder to other conditions (i.e., a pulled trigger). Yet, it is clearly not employed, in any serious way, as an information bearer or representational device. So it is clear that a structure's nomic dependency upon some condition can be relevant to its functional role, without the structure thereby serving as any sort of representation. As Dretske himself notes, it is possible for something to be a potential representation (i.e., possess states that *could* inform about other states of affairs) without the relevant information being exploited (without anything *actually* using those states to learn about those states of affairs). Actually to function as something like an informer or representation, a structure needs to be plugged into the right sort of system in the right sort of way. The relevant entailment relations need be put to a very specific sort of use. Being plugged into a system as simply a reliable causal mediator does not qualify.

Another way to see all of this is to consider that the conventional sort of causal dependencies or co-variation properties that take center stage in theories of representation are not the only types of relations that could be exploited to make something into a representation. Other types of law-like relations can be exploited as well. For example, if A is always larger than B, then, in the deflationary sense we are now using the term, A carries information about the size of B; that is, the size of A could be used to tell someone something about the size of B. If A is heavier than B, or if A is always within a certain distance of B, then the weight or position of A can serve to inform a cognitive agent about the weight or position of B. In all of these cases, specific types of law-like relations between two objects (larger than, heavier than, close to, etc.) can and sometimes are exploited by cognitive systems like ourselves such that our knowledge of the status of one of the objects can generate knowledge of the status of the other one as well. When this happens, one of the objects is serving as a type of representational device.

Keeping this in mind, we can see that there is a wide range of cases in which a state or structure possesses the sort of relational property that could be relevant for a representational role, but the state or structure in question is not playing a representational role, *even though that same relational property is essential to the role it does play.* If it is a law that A is always heavier than B, then this relational property of A (along with A's weight) *could* be used to learn something about the weight of B. If we learn that A's weight is 5 lbs, then this tells us that B weighs less than 5 lbs. Thus, this "nomically heavier than" relation makes A into a potential representation with regard to the weight of B. But this same "heavier than" property of A could be exploited in an entirely different way. It could be exploited so that A functions as a counter-weight for B. Or, it could be exploited so that A is recruited as an anchor for B. The "heavier than" feature of A could be explanatorily and causally relevant to a variety of different functional roles for A, not all (or even many) of which would be representational in nature. In other words, the following three claims are compatible:

(1) A is recruited for a task because of a certain relation it stands in to B.
(2) The relation A stands in to B is one that could allow A to a play a representational role with respect to B.
(3) The task A actually performs is *not* representational in nature – it is used for some job entirely non-representational.

All this brings us to the following observation: If we equate being an indicator with being a nomic dependent, then Dretske cannot establish that a structure is a representation by showing that it functions as an indicator because, trivially, functioning as an indicator just means functioning as a nomic dependent, and, as we've just seen, there are all sorts of ways to function as a nomic dependent without being a representation. The "nomically dependent upon" (or "reliably caused by" or "regularly co-varies with") relation can bestow a number of different functions upon a structure *other than* serving as some type of representation. In fact, in the sorts of cases and arrangements Dretske describes, the functional role of the indicator is most naturally described as a non-representational (but reliable) intermediary or relay switch. Just as with the realist interpretation of information, on the deflationary analysis, Dretske's account of how a state functions as an indicator fails to show how it serves as a representation. The job description challenge is thus unanswered.

4.4 FURTHER DIMENSIONS OF THE RECEPTOR NOTION

One obvious question worth asking at this stage is how do we tell the difference between the two sorts of cases we considered at the end of the last section? How do we demarcate between (a) cases where the nomic regularity is relevant to a non-representational function and (b) cases where it makes sense to say that nomic dependency helps something serve as a representation? What makes Dretske's examples of bi-metallic strips and magnetosomes cases of the former, as I've claimed, and not the latter?

One way to try to answer this question is by contrasting uncontroversial examples of mere causal mediation with uncontroversial cases of actual representation. Consider the difference between the firing pin in a gun and the mercury in a thermometer. In the case of the former, there is no sense in which the information carried by the firing pin is exploited in its normal functioning. While an extended pin is a sign of a pulled trigger (along with a host of other things), nothing in any sense interprets this sign or exploits the link to learn something new. By contrast, the mercury in a thermometer functions as a representation because there is a clear sense in which it serves to inform people who want to learn about the temperature. This suggests that the way to distinguish structures that are serving as relay switches or mere causal mediators, on the one hand, and structures that are serving as genuine representations, on the other, is that for the latter, but not the former, we have a process that is properly or naturally viewed as something like *learning about* or *making inferences about* some state of affairs with which the representation stands in some nomic relation. Perhaps one reason we mistakenly regard the bi-metallic strip in a thermostat or the magnetosomes in bacteria as representations (and not firing pins in guns or immune responses) is because they resemble *actual* representational devices – like thermometers and compasses – that, in their proper functioning, have this added dimension. Since this added element does not exist in thermostats and the bacteria – since there is no motivation for thinking the furnace or the flagellum are doing anything like acquiring knowledge or making inferences – we have no reason to view these and other receptor-like structures as serving as representations.

If to be used as representations, receptor-like structures require these more sophisticated processes (like learning or making inferences), then this would suggest they present a poor model for understanding the nature of mental representation. Recall that Peirce denied that representations of any form could function as representations without a full-blown interpreter. I've argued that he was wrong to claim this about his notion of icons. We can indeed develop a working notion of representation based upon the principles associated with models and maps without also invoking a sophisticated interpreting mind that uses them. In other words, the sort of representation that is based upon the isomorphism between a model and its target can be built into the workings of a purely mechanical system, despite the lack of a thinking interpreter. But it seems the same does not hold true for what Pierce called indices. Indices are (more or less) equivalent to what I've been calling receptors, and for something to qualify as a representation of this sort, it appears there needs to be a fairly sophisticated cognitive agent that employs such things – not internally (as part of its own cognitive machinery), but externally, as something on the basis of which inferences can be made and propositions can be learned. Bi-metallic strips play a representational role in thermometers, but not in thermostats. Peirce, it turns out, was wrong about icons but right about indices.

A potential worry about my analysis is that I'm employing a double-standard, using one set of criteria for defending the value of IO-representation and S-representation, but adopting much tougher standards for the receptor notion. In earlier chapters, we saw how misguided it would be to attack a representational theory by insisting that it is *possible* to describe any physical system in purely non-representational terms. Yet it might appear to some that I am doing precisely that in my criticism of the receptor notion. That is, it might appear that I am rejecting the idea that receptors are representations simply because the systems in which they appear can be given a purely causal-physical analysis.

But this worry is misplaced. The problem with receptor representations is not that the systems that allegedly use them *can be* given a purely causal-physical, non-representational characterization. Rather, the problem is that the theoretical frameworks that invoke them, despite various mis-characterizations, actually assign to them a non-representational role. Serving as a structure/state that is reliably triggered by some other condition, or that is recruited to cause something to happen whenever a specific condition obtains, is to play a role that, as such, is not representational in nature. There is nothing about this job description that suggests the role receptors play is based upon their having content, or because they stand for or stand in for, something else. By contrast, in the CCTC framework, there exist theoretical considerations and commitments, stemming from a specific explanatory strategy, that drive us to characterize certain states as playing a representational role – to regard them as employed *as* representations by the system. Consequently, the receptor notion of representation, unlike the IO notion and S notion, really *isn't* a representational notion at all. There are, no doubt, states that play the functional role cognitive scientists ascribe to receptors and that are even tuned to respond to very specific parameters and types of stimuli (see, for example, Sterelny 2003). The claim is not that we never find systems that operate in this manner. What is being denied is that such a state – whether it be a chemical-receptor, or photo-receptor, or a face cell, or what have you – should be treated as a type of representation. Sensory receptors are functionally similar to protein receptors on cell membranes. When the mechanics of cell membrane protein receptors are fully articulated, few people are inclined to claim that protein receptors actually serve as representations. Instead, they are seen as structures that reliably transport specific molecules (or other chemical or electrical phenomena) into the cell; they serve as a type of non-representational transducer[7]. Similarly, when the mechanics of receptors in our sensory and cognitive systems are properly understood, we see that they also play a relaying role, not a representational role.

In chapter 6, I will revisit the fairness of my analysis by offering a more direct comparison between the S-representation notion and receptor notion. I will also provide a more detailed analysis of why the former, but not the latter, works as a theoretical posit of psychology. For now, I want to explore a different matter about what my analysis of representation implies about the way we often think about representation in non-cognitive systems. Consider two sorts of mechanical devices. If you go into a modern public restroom, you are likely to encounter one of two types of faucets. The first is the old-fashioned sort where you simply turn a handle for water. The second type is the more sophisticated, though often frustrating, fully automated version (supposedly) designed to initiate water flow whenever a person puts her hand under the faucet. On a popular way of looking at things, there is a fundamental difference in the workings of the two sorts of faucets. One is viewed as a straightforward mechanical device, whereas the other is viewed as using a type of sensor that serves to represent the presence of someone's hands. Moreover, the distinction between the two sorts of faucets (and similar devices) is generally assumed to provide us with an important lesson about the nature of cognitive systems – particularly with regard to sensory systems. The automated system is thought to incorporate the same fundamental principles of actual perceptual systems, whereas the mechanical system is not.

An upshot of our discussion is that this distinction between these two sorts of devices is metaphysically untenable. Both sorts of faucets operate on the same basic principles and there is no sense in which one is a representational device and the other one isn't. Most automated faucets use infrared sensors that respond to photons in the thermal infrared spectrum, as these are emitted by living systems. The photons bump into the sensors and cause them to initiate a process that eventuates in water flow. While I acknowledge that the automated system can provide an understanding of how biological systems work (because both systems make use of photo-receptors), what I deny is that the *way* such systems work should be understood as fundamentally different from purely

[7] In engineering, a transducer is typically characterized as something that converts one form of energy into another. This strikes me as an appropriate job description for many instances of receptor representation. However, I have avoided the term "transducer" because it is often associated with semantically loaded notions such as "signal translation" or "sign interpretation".

mechanical, non-representational systems. Both faucets employ elements that are designed to operate in a way that exploits causal relations between states of the world and states of the system. While the implementation details are different, both use the same *sort of* functional components and processes. The faucet handle and the infrared sensor have the same basic role – they are supposed to mediate a causal link between the presence of people who want to wash their hands and the flow of water.

Of course, that is not to say that there aren't *any* differences between the two faucets. The point is that none of these differences are *intentionally relevant.*[8] For instance, in the case of the manual faucet, there needs to be physical contact with the user's hand and the handle. In the case of the automated faucet, the contact between the user's hand and the relevant component (the infrared sensor) is mediated by photons. But this hardly seems to matter. If we used some intermediary device, like a pole, to turn the faucet handle, this would not convert the manual faucet into a representational device. Another difference is that the infrared sensor is responsive to a narrower range of conditions than the manual faucet. Yet once again, this is not an intentionally relevant consideration. If we were to modify the handle so that only certain types of hands (say, really strong ones) could turn it on, no one would think this transforms the manual faucet into a representational system. Furthermore, different kinds of conditions might cause the faucets to malfunction; a janitor's mop might accidently turn the faucet's handle, while someone strolling past might accidentally trigger the automatic faucet's detector. Yet these different causes of inappropriate faucet activation don't change the fundamental nature of the malfunctioning – the difference doesn't make the latter into a case of misrepresentation. Finally, the actual workings of the manual faucet are more transparent to most of us than the mechanisms of the automated faucet. However, this is just a contingent fact about our understanding of devices. At most, this explains why we are more inclined to treat the two systems as different, not a justification for treating them as different.

Given the degree of overlap between the two sorts of systems, we should either say that they are both representational systems, or that neither are. If we adopt the former position, and claim that ordinary faucet handles actually do serve a representational function, then this would suggest that any mechanical device or system that employs structures that in their proper functional role are supposed to do something when and only

[8] The phrase "intentionally relevant" is meant to be analogous to the phrase "morally relevant" in ethics. It specifies properties that contribute to something's status as a representation.

when certain conditions obtain are actually representational systems. But practically every complex system employs structures of this sort. This would not only lead to a sort of pan-representationalism, it would also force us to abandon our ordinary understanding of what it means for something to function as a representation. From a theoretical standpoint, it would drastically deflate the Representational Theory of Mind into something like the "Responder to Specific Causes Theory of Mind." On the other hand, if we adopt the latter position, then we need to abandon the view that receptor-like mechanisms are representational. I believe that this is the sensible option to adopt.

Returning to cognitive theorizing, what should we say about the vector analyses of the hidden units of connectionist networks? As we noted above, many writers treat these units as serving as representations in part because of what vector analyses reveal about how the hidden units react to different stimuli. When we plot the activation patterns in vector space we find that there are intriguing groupings, with patterns from distal inputs (echoes from undersea mines and rocks, English vowels and consonants, etc.) clustered together in ways that reflect real-world similarities. Don't these clusters in vector space reveal that the receptor-like hidden units are serving to encode distinct chunks of information about the world? Aren't they evidence of a representational function?

It is important to be clear about what the vector analyses actually tell us. As I've previously noted (Ramsey 1997), the clusters tell us that the internal units of a network go into similar states when the network is presented with similar stimuli. They reveal a response profile in the network that suggests inputs from the same types of conditions are processed in similar ways. This is indeed informative about the way the networks solve various problems. But there is nothing about this fact that suggests the hidden units are playing a representational role. To draw the conclusion that a vector analysis reveals a network's representation scheme, you have to first assume the hidden units are, in fact, serving as representations. The mere clustering in the response profile doesn't show this. There are, after all, plenty of non-representational systems that admit of the same sort of vector clustering. If we do a similar analysis of a climber's blood at different altitudes (using variables like oxygen or nitrogen levels as vectors), we'll find similar clusters corresponding to different elevations, food intake, and so on. Despite this, no one thinks the climber's bloodstream is serving a representational function. The results of vector analyses, though intriguing and explanatorily useful, don't in any way reveal that the hidden unit activation patterns serve as representations.

4.5 DOES IT REALLY MATTER?

My aim has been to show what is wrong with the receptor notion and why it is mistaken to view these sorts of structures as representations. However, in response to these arguments, people often complain that I am simply being stingy with the word "representation." After all, why not treat receptor states as some low-level type of representation. Perhaps there are no deep differences between things like firing pins or immune responses on the one hand, and states we call receptor representations on the other. But what is the harm in referring to the latter as representational states? If doing so is a conceptual confusion, then it appears to be a conceptual confusion that has no negative consequences.

It is often unclear to me how to understand this response. Taken as a point about language and our ability to call things what we want, then it strikes me as correct, but somewhat silly. We can indeed choose to use words however we like, and so we can call structures that have no representational function "representations," just as we can choose to call hearts "kidneys" or dogs "cats." But I believe the complaint is based upon more serious considerations than this. The more serious consideration is not so much about our use of words, but instead stems from a lingering pull toward the "intentional stance" with regard to receptor-type systems. I must confess that I also feel this pull to some degree. If we *are* so inclined, and nothing is really at stake by adopting this stance, then why not go ahead and do so?

The answer is that, as I pointed out in chapter 1, there actually is a fair bit at stake here. Even if it were true that cognitive research has not been severely undermined by the misconception of receptors as representations, this misconception has, if nothing else, contributed to the current confusion in the field about the nature of representation. In the cognitive neurosciences, the terms "representation" and "information" are increasingly used in ways that seem to mean little more than "reactive neuron" or "causal activity." This only serves to blur a useful distinction between representational and non-representational mechanisms of the mind. But moreover, there are a variety of ways in which invoking of the receptor notion can derail cognitive research.

When we treat various elements of cognitive systems as representations, we commit ourselves to a certain model or picture of the system's functional architecture. If we regard certain things as playing a role that they are not actually playing, then that picture is inevitably going to be mistaken. The problem with receptor representation in cognitive explanation is similar to the following scenario: Suppose you are ignorant of how a car engine works, so you investigate the operations of its various components. You notice that the firing rate of a series of units (which in fact are spark-plugs) nomically depends upon the depression of the car's accelerator pedal. Invoking the receptor notion, you thereby conclude that the function of these devices is to represent the position of the accelerator pedal, which in turn leads down a certain research path. For example, you devote time and energy to investigating why the pedal representations are located where they are in the engine. You also speculate about which aspects of the pedal are actually supposed to be represented by these units (is it really the position of the pedal, or is it instead the pressure placed on the pedal?) You also investigate the consequences of misrepresentation (when the pedal becomes broken) and you also explore hypotheses about why the information is encoded with similar representations (the other spark plugs) that fire in an alternating pattern. You wonder if there is additional information that is being relayed through this firing pattern, and if so what that information could be. Of course, all of this work leads you away from seeing the spark-plugs as ignition devices that are controlled by the accelerator pedal and cause a combustion sequence that generates in a spinning drive-shaft. Generally, if you assume that something plays a role that it actually doesn't play, then your functional analysis of the system is going to be off the mark.

All of this may seem fanciful, but there are actual cases where the receptor notion has undermined research in just this manner. For example, in a short but important article, neuroscientists Walter Freeman and Christine Skarda (1990) describe how their research on the olfactory system was derailed for several years because they mistakenly assumed that certain neural arrays functioned to represent olfactory stimuli because of the way they responded to that stimuli. This interpretation led them away from a more accurate understanding of the roles these neural states play in brain dynamics. They found that "thinking of brain function in terms of representation seriously impedes progress toward genuine understanding" (1990, p. 376). Their analysis is worth quoting at length:

> We have found that thinking of brain function in terms of representation seriously impedes progress toward genuine understanding . . . The EEGs of the olfactory bulb and cortex show a brief oscillatory burst of potential that accompanies each inhalation . . . We have shown that a stereotypical pattern recurs whenever a particular odorant is presented that the animal has been trained to respond to. For more than 10 years we tried to say . . . that each burst served to represent the odorant with which we correlated it, and that the pattern was like a search image that served to symbolize the presence or absence of the odorant that the system was looking for. But such interpretations were misleading. They encouraged us to view

neural activity as a function of the features and causal impact of stimuli on the organism and to look for a reflection of the environment within by correlating features of the stimuli with neural activity. This was a mistake. After years of sifting through our data, we identified the problem: it was the concept of representation . . . Once we stopped looking at neural activity patterns as representations of odorants, we began to ask a new set of questions. Instead of focusing on pattern invariance and storage capacity, we began to ask how these patterns could be generated in the first place from less ordered initial conditions. What are the temporal dynamics of their development and evolution? What are their effects on the neurons to which they transmit? What kinds of structural changes in brains do they require and do they lead to? What neuromodulators do these pattern changes require? What principles of neural operations do these dynamical processes exemplify and instantiate? In short, we began to focus less on the outside world that is being put into the brain and more on what brains are doing. (1990, pp. 376–377)

Freeman and Skarda go on to describe how addressing these questions and adopting this different perspective allowed them to develop explanations of neural dynamics that did not appeal to representations. In their new account, the brain is viewed as more of a self-organizing system, with input that serves not to inform but to trigger chaotic patterns of activity out of which new states of organization can emerge. Variations in the burst that were originally thought to be insignificant noise in a represented signal were now seen as important sources of these necessary patterns of activity. They go on to conclude that invoking representations will "impede further advances toward our goal of understanding brain function, because they deflect us from the hard problems of determining what neurons do . . ." (p. 379).

Given my defense of the IO-representation and the S-representation, I'm not prepared to go as far as Freeman and Skarda and issue a blanket rejection of all notions of representation from our understanding of brain function. Nor am I in a position to claim that their non-representational theory of olfactory processing is correct. Nonetheless, I do believe their testimonial provides a good example of how the receptor notion can undermine scientific progress by promoting a misguided analysis of the functional role of internal cognitive states. It is not just that it is explanatorily unnecessary to treat receptor-type states as playing a representational role. It is also potentially damaging to a more accurate understanding of brain dynamics.

4.6 SUMMARY

The principal claim of this chapter is, from the standpoint of received wisdom, fairly radical. I've argued that one of the most popular ways of thinking about representation in cognitive science is confused and ought to be discontinued. Despite its common appeal, the receptor notion of representation comes with a job description that, in this context, has little to do with the role of representing. Instead, it involves the role of becoming active when and only when some specific condition obtains. Such a property certainly *can* be relevant to something's serving as a type of representation in the right sort of circumstances. But those circumstances appear to require an independent cognitive agent that exploits this feature to learn something new. When we look at the role of receptors *inside* of cognitive systems, as described by cognitive theories that employ them, we see that the role is better described as something like a reliable causal mediator or relay circuit which, as such, is not representational in nature. In other words, when a causal/physical system (like the brain) is described as performing various cognitive tasks by employing a structure that has the job of causing something to occur when and only when something else occurs, then the system is not, on the basis of this description alone, employing internal representations.

In an effort to supplement the receptor notion and see if it can be enhanced to meet the job description challenge, we looked at the account of cognitive representation offered by Dretske. The virtue of Dretske's account, besides its sophistication, is that it involves the same basic principles at work in the receptor notion and Dretske frames his analysis in exactly the right way, namely, by focusing on what is involved in something *functioning* as a representation. Yet while Dretske asks the right questions, his account does not provide suitable answers. Dretske suggests that receptor-type states qualify as representations by virtue of the way in which they are incorporated into the cognitive architecture. He suggests that if receptors are recruited because of the way they respond to certain distal conditions, then they are representations. Yet this doesn't help us with our original concern. The structures Dretske describes do intuitively serve a functional role, but the functional role is still that of a reliable causal mediator, and not that of a representation. Of course, there might be other ways in which the receptor notion can be revised or re-described so that it can be seen to be an explanatorily valuable notion of representation; we will look at some of these in chapter 6. As we will see, these other accounts fare no better than Dretske's.

The cognitive theories that invoke the receptor notion – particularly those in connectionism and the cognitive neurosciences – are often regarded as dramatic departures from the framework of the CCTC. But it turns out that this perspective doesn't go far enough. These theories are a much greater departure from classical AI than generally appreciated. If the positing of receptor-style states is supposed to make these accounts *representational*, then the correct analysis is that, as it turns out, these aren't representational theories after all. If these theories turn out to be the right way to think about cognition, and if the only notion of representation they employ is the receptor notion, then it will turn out that the Representational Theory of Mind (RTM) is simply false. The so-called "Cognitive Revolution," which is generally thought to be based in large measure upon the invoking of inner representations, would be much less of a revolution than most commentators assume, a point we will return to in the final chapter.

Of course, many would deny that the receptor notion is the *only* notion of representation that appears in the newer accounts of cognition. Many would claim that there is another notion that has less to do with receptors and more to do with the *effects* of the state or structure in question. Thus, before we can declare that representationalism is in trouble in the contemporary cognitive neurosciences, we need to look at this other representational notion that is growing in popularity. The next chapter will examine this notion of representation – one that is based upon the idea that information can be "tacitly" encoded in structures and states that are distributed throughout the processing elements. Yet, as we'll see, this tacit notion also runs into trouble regarding the job description challenge. Like the receptor notion, the tacit notion generates more conceptual confusion than proper understanding of how the brain might work.

5

Tacit representation and its problems

In the last chapter we looked at one popular way of thinking about representation in many non-classical accounts of cognition, and saw how that notion of representation ran into serious difficulties. If I'm right, the receptor notion should be abandoned by cognitive scientists. While it may be beneficial to advance and adopt theories that posit internal relay structures, there is no explanatory benefit in viewing things that function in this manner as representations. Indeed, as we saw at the end of the last chapter, there are reasons to think that viewing them as representations leads to various theoretical problems. In this chapter, we will explore another, quite different family of representational notions that I will argue deserves a similar verdict. I'll refer to this family of representations as "tacit" representations. Because it involves a diverse range of structures, the tacit notion is not easily defined. Yet we can initially characterize it as based on the idea that there is a close link between the dispositional properties of a cognitive system, and a type of knowledge or information that the system represents "inexplicitly" or "tacitly." This characterization may still seem vague, but it will hopefully become clearer as we progress. The tacit notion is not a new notion of representation, but it has gained considerable significance in recent years with the advent of various non-classical accounts of cognition, especially connectionism. With this type of representation, it is typically assumed that there is no one-to-one correspondence between individual structures or states and specific items of content. Instead, large chunks of the system's architecture are assumed to represent a wide range of things, partly because of its various capacities. Recall that with the receptor notion of representation, structures are thought to represent in large measure because of the way they are causally influenced by certain conditions. With tacit representation, the causal order is reversed – structures are alleged to serve as representations, at least in part, because of their *effects*.

My aim is to argue that when we look closely at the tacit notion with regard to the job description challenge, we can see that like the receptor notion, it fails to meet that challenge. More specifically, I'll argue that the standard reasons to treat the functional architecture of cognitive systems as playing a tacit representational role are in fact no good, and that when people appeal to tacit representations they are often talking about nothing other than straightforward, non-representational dispositional states. Since there is no explanatory benefit in adding a representational or informational gloss to these dispositional states, I'll argue that the notion of tacit representation is metaphysically and explanatorily bankrupt and should be dropped altogether from our understanding of cognitive systems.

To show this, the chapter will have the following organization. In the next section, I'll try to clarify what the tacit notion entails by looking at four different ways common sense implies that people can believe something in a way that is implicit or tacit. I'll claim that only one of these captures the basic idea behind the tacit notion in cognitive science I want to look at. In Section 5.2, I'll offer a further explication of the tacit notion by illustrating a few of the ways it shows up in scientific theories as well as in philosophy. Then, in Section 5.3, I'll offer my critical evaluation of the tacit notion. After presenting what I regard as a popular line of reasoning in support of the tacit notion of representation, I'll argue that this line of reasoning is severely flawed, and that the standard motivations for appealing to the tacit notion of representation are equally flawed. I'll contend that when people appeal to tacit representation (or tacit knowledge) within cognitive systems, they are usually just talking about the dispositional nature of the system's architecture. Since there is no reason for thinking the dispositional properties of a cognitive system play any sort of representational role, there is no good reason to treat them as representational states. Section 5.4 offers a brief summary.

5.1 THE TACIT NOTION: COMMONSENSE ROOTS

It would help us to get a handle on the tacit notion if we could look at a few non-mental instances of the relevant representational principles, much as we have done with other notions of representation. Unfortunately, unlike the S-representation notion or the receptor notion, it is difficult to find examples of tacit representation in the non-mental realm. Holographic representations may offer one non-mental example of tacit representation, though it is actually far from clear that holograms encode information in the relevant sense. In fact, our basic understanding of tacit representation seems to be derived more from commonsense psychology than from our use of non-mental entities. Indeed, one possible motivation for invoking a similar notion in cognitive theories is to capture this element of commonsense psychology. We can therefore begin to get a handle on the tacit notion in cognitive science by looking at the different ways something similar pops up in folk psychology.

In truth, there are several different notions of folk mental representation and belief apart from the paradigm case of having an explicit, conscious thought. Here are four ways of believing or knowing something that might be (and have been) considered, in some sense, implicit or tacit:

(1) A belief that is unconscious (as opposed to conscious).
(2) A belief that is stored in memory (as opposed to currently active).
(3) A belief that is implied or entailed by other things believed (as opposed to being directly represented in some way).
(4) A belief that is somehow implicitly embodied throughout the agent's cognitive machinery (as opposed to being encoded by a discrete, identifiable state).

The first distinction is one that has worked its way into folk psychology, perhaps to some degree through the popularity of Freudian psychology. Often when people discuss subconscious or unconscious beliefs and desires, they are referring to states assumed to function much like normal mental states, playing many of the same causal roles, except these states have somehow stayed below the surface of conscious awareness. The "spotlight" of conscious introspection (or whatever metaphor you prefer) is not shining on them, even though they are still assumed to be associated with particular, concrete states. Thoughts of this nature may be conscious on other occasions, or perhaps they have never been conscious and could only become conscious through therapy or some other method. From a cognitive science perspective, such representational states can be regarded as similar to other representations, only they lack whatever extra feature is responsible for making a mental state conscious. In the psychological literature, unconscious information processing is often referred to as "implicit cognition" or "implicit information processing."

The second notion is based on the observation that we clearly have stored memories of things we have consciously and actively thought in the past. It is presumably accurate to say that five minutes ago, you possessed the belief that Rembrandt was a great artist even though this belief played no role in whatever cognitive processes were taking place in your head at the time. One natural way to think about such mental states is to assume they are the same basic representational structures that can later play an active role in our reasoning, but for the time being they are in a dormant mode, perhaps located in some sort of memory register. While stored memories are presumably not conscious, they are not the same thing as the unconscious mental states discussed in the last paragraph because they are not currently implicated in any mental operations. However, they share with unconscious thoughts an explicit form – that is, commonsense allows that they might be instantiated by fully discrete, identifiable representational structures.

The third notion stems from the practice of attributing beliefs to people even when they have never actually entertained such beliefs, either consciously or unconsciously. These beliefs are instead attributed because they are directly implied by things the believer actually does explicitly hold. I know John knows that the number of days in the week is seven. I may not think that John has ever consciously or unconsciously thought to himself that the number of days in the week is less than twelve. Nonetheless, I may say to someone (perhaps in defense of John), "Look, I know that John knows that the number of days in the week is less than twelve." We sometimes appear to ascribe views to people that we feel confident they would immediately accept if brought to their attention, even though we assume this has not yet happened. Following the writings of Dennett (1982), this is also sometimes referred to as "implicit representation" or "implicit belief." Implicitly held beliefs, in this sense, are beliefs that would be tokened if only certain inferences were made. In truth, it is unclear whether the ascription of this sort of implicit belief is actually the ascription of a current mental state, or rather just the ascription of a possible future belief. That is, when I say "John believes the number of days in a week is less than twelve," this may just be a misleading way of saying "John would immediately form the belief (which he has not yet formed and thus does not actually possess) that the number of days in the week is less than twelve if asked to consider the matter." Ascriptions of this sort may really be ascriptions of psychological dispositions – a tendency to form certain mental representations if certain conditions obtain.[1] As far as I'm aware, no studies have been done to reveal which reading of commonsense psychology is correct.

By contrast, the fourth and final notion of tacit belief is one where it is pretty clear that commonsense psychology ascribes a mental state assumed to actually, presently exist. This notion is rooted in the idea that there is a type of knowledge stored in whatever mental machinery underlying various cognitive skills and capacities, embodying one form of knowing *how* to do something. For example, we might say that someone

[1] See also Lycan (1986) for an interesting critical discussion.

knows how to ride a bicycle and assume that the person represents this knowledge in a way that does not involve any explicitly held beliefs with bike-riding content. We might think this sort of know-how is based on the general capacity of the individual's mind/brain – its ability to coordinate the body in the right way when on a bike, keeping the agent upright and moving forward. Presumably, this bike-riding ability is grounded in some sort of information encoding; after all, we *learn* to ride a bike. Yet folk psychology seems to allow that the encoding is such that no discrete individual representations are tokened with specifiable propositional content. It allows the information to be tacitly represented in the mental architecture that is causally responsible for one's bike-riding skills.

All four of these commonsense notions of belief or knowledge have analogues in the cognitive science literature. Cognitive scientists have posited unconscious mental representations, stored memories, dispositions to make trivial inferences, and a sort of tacit "know-how." However, the first three notions do not describe truly unique *forms* of representation. In fact, all three of these sorts of believing could be instantiated by versions of the types of representational notions we have already discussed in earlier chapters. For example, unconscious beliefs could be implemented by unconscious S-representations, or memories could simply be stored elements of a model. Thus, it is possible that structures implementing the first three notions of implicit belief could actually serve as representations (and thus answer the job description challenge) in the manner described in chapter 3, despite being unconscious or stored or merely implied by other beliefs. The first three notions do not capture a fundamentally distinctive *way* that structures or states might serve as representations.

The fourth notion, however, does present the folk analogue of (and perhaps inspiration for) a distinctive, qualitatively different notion of representation that appears in cognitive science theorizing. This way of thinking about representation has a long history in psychology and artificial intelligence, but it has taken center stage in newer non-classical theories, especially in connectionist accounts of the mind. Like its common sense analogue, it is based on the assumption that the problem-solving machinery of a cognitive system encodes a type of knowledge that the system exploits when performing cognitive tasks. Thus, the notion is closely connected to the functional architecture of the system in question. It deserves to be treated as a distinctive notion of representation because the *way* the structures are thought to serve as representation is not captured by any of the earlier notions we have discussed. The functional architecture of a system (or at the least the relevant aspect of that architecture) is not thought to play a representational role because it serves as computational inputs or outputs, or because it is thought to serve as elements of a model or simulation, or even because it is nomically correlated with environmental stimuli. Instead, components of the system are thought to play a representational role in this sense because of their potential to generate the right sort of processes and input–output mappings. It would not be too far off the mark to characterize this notion as a "dispositional" notion of representation. It is this sort of representational concept that the remainder of this chapter will examine.

5.2 TACIT REPRESENTATION IN SCIENCE AND PHILOSOPHY

5.2.1 Connectionism and cognitive neuroscience

To get a better handle on tacit representation, it will help to briefly look at some of the theoretical frameworks in which it appears. Connectionist modeling – especially the sort that involves feed-forward networks – is where the notion of tacit representation under consideration has become most prevalent. We have already discussed how the pattern of activity of the internal "hidden" units of a network are generally thought to serve as receptor-type representations. These are often characterized as "distributed representations" because the pattern of activity is distributed over the same individual units. At the same time, these distinct activation patterns are fully discrete states that, were they to play a representational role, would be a form of explicit representation. Yet there is another sense of distributed representation associated with connectionism that is not based on activity patterns of the internal units. Instead, it is associated with the weighted connections between the individual units. These connections – inspired by the axonal and dendritic connections between neurons in the brain – are responsible for transmitting excitatory and inhibitory causation between individual nodes in a network. Often, networks of this sort are developed through a training or "learning" phase during which the connection weights are adjusted both in terms of their excitatory or inhibitory nature and in terms of the strength of the "signal" they transmit. For example, after being adjusted, a connection between an input node and a hidden node might be characterized as, say, "+65," which would mean that the activity of the input node would trigger activation in the hidden node to a degree of 65 (where "65" indicates some pre-determined level of strength). Once the network responds to inputs in an appropriate manner, the network's connection weights remain fixed and, depending on how well the system generalizes to new cases, the network is regarded as having acquired the new problem-solving skill (Smolensky 1988; Bechtel and Abrahamsen 2001).

It is within these connections that neural networks are thought to embody a form of tacit representation. Their representational status is "tacit" in that no single weighted connection is thought to correspond to any single bit of information. Instead, the encoded content – assumed to be either propositional or some form of non-propositional content – is typically thought to be both (a) at least partially spread out over the bulk of different connections and (b) "superposed," so that the same weight configuration serves to represent different items of knowledge. As McClelland and Rumelhart put it, "each memory trace is distributed over many different connections, and each connection participates in many different memory traces. The traces of different mental states are therefore superimposed in the same set of weights" (Rumelhart and McClelland 1986b, p. 176). Thus, the smallest processing unit that lends itself to content ascription is thought to be the entire connection matrix, which is characterized as encoding the entirety of the system's stored information. For example, if the connections are thought to encode some set of propositions, then the most fine-grained semantic analysis that is possible is a holistic one that treats the entire connection configuration as representing the entire set of propositions. A more fine-grained analysis is not possible. For most connectionist networks of this type, "almost all knowledge is *implicit* in the structure of the device that carries out the task rather than *explicit* in the states of units themselves" (Rumelhart and McClelland 1986a, p. 75).

The tacit nature of this alleged mode of information storage is generally regarded as both a plus and minus by cognitive researchers using connectionist models. On the one hand, the distributed nature of the representation is thought to allow for a feature of real brains known as "graceful degradation." Since no single element represents a specific piece of content, and information is supposedly distributed over many connection weights, individual connections can break down without the loss of particular representations. Many connectionist modelers regard this as a biologically plausible aspect of neural networks, since real synapses deteriorate all the time without the loss of specific memories or elements of knowledge. The connectionist framework is also thought to offer a more biologically plausible account of learning. The Hebbian idea that learning is achieved by altering the strength of excitatory and inhibitory links between neurons corresponds nicely with the idea that information is acquired and stored in networks through the modification of connection weights. Yet despite these advantages, the tacit representational form also has its drawbacks. Rumelhart and McClelland point out that "{e}ach connection strength is involved in storing many patterns, so it is impossible to point to a particular place where the memory for a particular item is stored" (1986a, p. 80). This makes it extremely difficult to "de-bug" a network that is not performing as hoped. Because the system's entire knowledge base is thought to supervene concurrently on the same static weight configuration, system modification alters all of the alleged stored knowledge and does not specifically target only faulty stored pieces of information.

One of the most unique and intriguing aspects of this way of thinking about information encoding is the idea of superpositional representation – that distinct elements of content are actually simultaneously represented by the exact same underlying configuration. In conventional computers, distinct representations are stored in distinct items located in a register that typically functions like a file cabinet. Representational symbols are stored away and then later retrieved by using something akin to distinct addresses. By contrast, networks employ the single static weight configuration to play the same role as a storage device, even though the connection configuration has no distinct "place" or address for the different bits of information. As Clark puts it, "[t]wo representations are fully superposed if the resources used to represent item 1 are coextensive with those used to represent item 2" (Clark 1993, p. 17). With superpositional representation, it's as if we had a functioning file cabinet but with only a single basic file storing all pieces of data under the same address without any representational elements standing for particular pieces of data. As might be expected, precisely how to describe the true nature of superpositional representation, and the actual procedure by which individual contents are brought to bear on the processing, is far from clear. For example, it is unclear whether there are actually several different representations, each embodied in a single weight configuration, or whether there is only one representation that somehow manages to represent many different things.[2] Nevertheless, superpositional representation has become an accepted way to think about stored information in connectionist networks.

A prototypical connectionist model that appeals to the notion of tacit representation has recently been presented by Rogers and McClelland (2004). Their model, which is an extension of an earlier network developed by Rumelhart (1990), is designed to explain the cognitive machinery responsible for the way we make categorization judgments and assign

[2] See van Gelder (1991).

various properties to individual things, something the modelers describe as "semantic cognition." The model involves a number of explicit representations, with one set of input nodes where each unit designates items (like canaries), another set of input nodes that represent various abstract relations (like capacities), and individual output units that designate attributes (like singing). The network also involves two layers of hidden units that mediate between the input and output sets of units. As Rogers and McClelland report, the network accounts for a number of explananda associated with conceptual development and performance, including the way conceptual development typically moves from broad conceptual divisions to more fine-grained distinctions, how this process is reversed in dementia, why it is that some properties are learned more rapidly than others, why preverbal infants are more responsive to abstract (rather than immediately observed) properties of objects, and so forth for an impressive collection of conceptual phenomena. What matters for our discussion is what the model-builders say about the network's acquired and stored knowledge in the weights. As they note, "[a] key aspect of our theory of conceptual knowledge acquisition, captured in the Rumelhart model, is the idea that it occurs through the very gradual adaptation of the connection weights that underlie the processing and representation of semantic content, driven by the overall structure of experience" (Rogers and McClelland 2004, p. 63). The authors call this sort of resulting knowledge "inchoate," which means "the knowledge is built into the apparatus that carries out the process of generating outputs, such that it causes people to behave in certain ways, but is expressly not available for readout as such" (2004, p. 329). In other words, with this account, our stored conceptual knowledge is tacitly represented in cognitive machinery, with no specific element serving to represent any particular item.

At a lower level of analysis, in the computational neurosciences, we also find a strong appeal to the tacit notion. The general idea that information is stored in the brain in a tacit, distributed manner is not new to the neurosciences, and can be found in Lashley's account of memory where he claimed that, "it is not possible to demonstrate the isolated localization of a memory trace anywhere within the nervous system . . . The engram is represented throughout the area" (Lashley 1960, pp. 501–502). Yet because connectionist modeling has had an enormous influence on the cognitive and computational neurosciences, especially with regard to their accounts of learning, memory, and information processing, the notion of tacit representation has become much more prominent in recent years in the brain sciences. Connectionist-inspired cognitive modeling has brought to the forefront the idea that neural structures and synapses represent superposed chunks of information in this tacit manner, and the field of cognitive neuroscience has produced several theories of neural learning and memory that simply adopt the connectionist perspective on tacit representation. Indeed, this is increasingly becoming a popular way of treating the computational role of actual synaptic junctures between dendrites and axons. For example, in their discussion of theories of information processing and learning in the hippocampus, Shapiro and Eichenbaum state, "the relationships among identified stimuli are stored as sets of synaptic weights on groups of single cells in the hippocampus . . . an hierarchical organization of relational representations is encoded by recurrent connections among relational cells" (1997, p. 118). Similarly, theories about the acquisition of different cognitive and motor skills in various regions of the brain often adopt a connectionist account of acquired representations through the modification of synaptic links (Churchland and Sejonowski 1992; McNaughton 1989).

5.2.2 Tacit representation in the CCTC

So far, I have suggested that tacit representation is exclusively found in the newer, non-classical accounts of information storage and processing. But this is somewhat misleading. While the notion of tacit representation has taken on much greater significance in non-classical theories of cognition, it also has a history in classical computational models. The CCTC framework is primarily committed to explicit representations of the sort discussed in chapter 3, but traditional AI researchers have also claimed that there is a kind of information storage to be found not just in these explicit data structures, but also in the functional architecture of the computational system itself. The idea is wonderfully captured by Dennett in the following, now-famous, passage:

> In a recent conversation with the designer of a chess-playing program I heard the following criticism of a rival program: "It thinks it should get its queen out early." This ascribes a propositional attitude to the program in a very useful and predictive way, for as the designer went on to say, one can usually count on chasing that queen around the board. But for all the main levels of explicit representation to be found in that program, nowhere is anything roughly synonymous with "I should get my queen out early" explicitly tokened. The level of analysis to which the designer's remark belongs describes features of the program that are, in an entirely innocent way, emergent properties of the computational processes that have "engineering reality". I see no reason to believe that the relation between belief-talk and psychological talk will be any more direct. (Dennett, 1978, p. 107)

While Dennett's main point is about the nature of our ordinary notion of belief, he is also endorsing the idea that the CCTC makes use of a valuable notion of non-explicit representation – a notion of representation based on the functional dynamics of classical computational systems. Dennett's point is that it is perfectly legitimate and useful to ascribe representations to the system even though there is no identifiable structure or state that has the *exclusive* role of serving as the representation in question. Dennett is not alone in the perspective. Endorsing this view, Clapin claims that "a powerful conceptual tool in cognitive science" is the idea that "functional architectures represent. They represent tacitly, and this tacit content is distinct from whatever is represented explicitly by the symbols that are supported by the functional architecture" (Clapin 2002, p. 299). In their defense of tacit representation, both Dennett and Clapin resurrect a well-known argument of Ryle's against what Ryle called the "intellectualist legend" (Ryle 1949). The "legend" is, by and large, what we now describe as the Representational Theory of Mind, discussed earlier in chapter 2 – the view that we can account for mental processes by positing inner representations. As these authors interpret Ryle, he argued that any system with internal representations would need sophisticated internal representation *users* – mental sub-systems that must have considerable sophistication and, in fact, a certain sort of knowledge. Thus, these internal sub-systems must also possess representations for this knowledge, the use of which would require yet deeper representation users. Representationalism, then, appears to lead to a sort of regress that we have discussed in previous chapters. Ryle's conclusion is that representationalism is a faulty framework for understanding mental processes.

To handle this regress worry, Dennett and Clapin suggest that AI researchers who invoke a notion of explicit representation are also thereby forced to invoke a notion of tacit representation. Tacit representations provide the sort of know-how that is required for using explicit representations. This know-how is implicitly embodied in the dispositional nature of the system's architecture, and because this knowledge is tacitly represented, there is, apparently, no further need for an inner homunculus to interpret or use it; thus, the regress is avoided. In other words, if representational content is only tacitly encoded, then, it is assumed, this content can be exploited in a way that does not require the sort of sophisticated mentalistic processes that worried Ryle. As Clapin claims, "tacit content is required for explicit content, and thus there will be no full story about mental content without a proper consideration of tacit content" (Clapin 2002, p. 295).

In discussions of classical AI, the notion of tacit representation is sometimes linked to two other important distinctions that appear throughout the literature. The first is the distinction between "declarative" and "procedural" knowledge or memory (Anderson 2000; Brachman and Levesque 2004). The former is generally assumed to be explicitly represented, often in propositional form. It is frequently characterized as the sort of knowledge that we might access for linguistic expression. An example would be a person's belief that the capital of Oregon is in Salem. By contrast, procedural knowledge is generally characterized as the embodiment of a cognitive system's know-how, and is often assumed to be not directly accessible to consciousness or verbal reports. An example of this sort of knowledge would be the bike-riding know-how we discussed earlier. Procedural knowledge is sometimes thought to be represented in non-propositional form, and in some accounts is embodied in the dispositional properties of the system. Thus, the distinction between declarative and procedural representation often corresponds directly to the distinction between explicit and tacit representation. However, it should be noted that in many CCTC models, the data structures that comprise the system's know-how are actually explicitly stored and tokened.

The second important distinction is between computational systems or sub-systems that are characterized as "rule-governed" and those that are characterized as "rule-describable" (Cummins 1983; Clark 1991). Rule-governed systems are generally assumed to possess, in some way, explicit and causally salient representations of commands or procedures. As we noted in chapter 3, in many cases there are reasons to be skeptical about the claim that these structures actually function as representational elements. But regardless of how we view them, they are nevertheless explicit, discrete structures that play specific causal roles. By contrast, systems are characterized as merely rule-describable if they behave in a manner consistent with certain principles expressed through rules or laws, yet no such rule is actually represented in the system or serves to causally influence its activities. Famously, a planet whose trajectory accords with Kepler's laws is merely rule describable in this sense; similarly, a pocket calculator may follow the rules of arithmetic without the rules being in any way represented in the calculator. Yet with the calculator, some (like Dennett) might say that the rules actually are tacitly represented within the calculator's inner workings. If we adopt this perspective, then it would appear that any distinction between computational systems that are "merely" rule describable without the rules being represented, and systems that do employ tacit representations of rules would evaporate. After all, any functional architecture whose operations accord with rule-like commands or propositions (e.g., "get the queen out early") can be said to tacitly represent those rule-like commands or propositions. Thus, one possible consequence of taking seriously the notion of tacit representation would be the need to abandon a traditional distinction between these two types of rule-following.

5.2.3 *Tacit representation in philosophy*

As with most of the notions of representation explored in earlier chapters, the tacit notion has a long history in philosophy. It goes back at least as far as Plato, who employed the Greek term "techne" to distinguish a kind of know-how that is not truth-evaluable. In more recent times, in the earlier half of the twentieth century, Ryle, Quine and other philosophers treated talk about beliefs and desires as disguised talk about multi-tracked dispositions. It is sometimes supposed that these philosophers were thereby endorsing a notion of tacit information storage, insofar as the dispositions they invoked are similar to what we've seen today characterized as tacit representations. Yet, this would be the wrong way to read these authors. Rather than putting forward a tacit notion of representation, they were actually reinterpreting talk that appeared to be about inner representations as actually about non-representational dispositions. Ryle and other philosophical behaviorists were anti-realists about inner representational states, tacit or otherwise, and maintained that mental talk is really about non-representational dispositions.

By contrast, in contemporary philosophy, there are a number of different advocates of the tacit notion, or something very close to it. Perhaps the best-known example of such a philosopher is Dennett (1978, 1987). Dennett's philosophy of mind is subtle and complex and not easily summarized, though for our purposes we can perhaps capture the basic idea. As we saw in chapter 3, Dennett claims that when we ascribe beliefs, desires and other propositional attitudes to agents, we do not actually assume that there are discrete, explicit representations that correspond with such ascriptions. Instead, we have adopted a heuristic predictive strategy – the "intentional stance" – in which belief-desire talk is used to capture patterns of the behavior of rational agents. According to this view, mental representations should be treated as abstracta, like centers of gravity. Although real, they are not concrete entities or states that can be somehow isolated and picked out inside the brain. According to Dennett, for a system, to be a true believer is to be "an *intentional system*, a system whose behavior is reliably and voluminously predictable via the intentional strategy" (1987, p. 15). What is it to adopt the intentional strategy? "[F]irst you decide to treat the object whose behavior is to be predicted as a rational agent; then you figure out what beliefs that agent ought to have, given its place in the world and its purpose. Then you figure out what desires it ought to have on the same considerations, and finally you predict that this rational agent will act to further its goals in the light of its beliefs" (1987, p. 17). A system with mental representation is thus a system whose behavior can be "reliably and voluminously" predicted in this way. Dennett unabashedly acknowledges that this might include systems we don't normally think of as mental, like coke machines and thermostats. For him, the possession of internal representations is not due to any sort of inner structure playing a representational sort of role. As he puts it:

> There need not, and cannot, be a separately specifiable state of the mechanical elements for each of the myriad intentional ascriptions, and thus it will not in many cases be possible to isolate any feature of the system at any level of abstraction and say, "This and just this is the feature in the design of this system responsible for those aspects of its behavior in virtue of which we ascribe to it the belief that P." (1978, p. 21)

While it is fairly clear what Dennett thinks beliefs are *not*, his appeal to abstracta and centers of gravity as models for understanding the nature of mental representation has not served to remove an air of mystery and uncertainty surrounding his positive view. Whether Dennett's account makes him a realist or some form of instrumentalist or even eliminativist with regard to beliefs is something that even Dennett himself has, at times, appeared unsure about. Yet one plausible reading of Dennett is to see him as a successor to earlier versions of dispositionalism. This reading is suggested by many of Dennett's writings, such as the example discussed earlier of the chess-playing system. While Dennett's view is clearly more sophisticated than Ryle's and other behaviorists, a natural way to interpret his various claims – such as his denial that representations correspond to a specifiable inner state, or that they stem from a predictive strategy – is to see him as endorsing the view that the possession of mental representations amounts to tacit information storage located in dispositional properties of a cognitive system. One author who not only reads Dennett this way, but also adopts such a view is Robert Stalnaker (1984). Stalnaker characterizes Dennett's position this way:

> Belief and desire, the strategy suggests, are correlative dispositional states of a potentially rational agent. To desire that P is to be disposed to act in ways that

would tend to bring it about that P in a world in which one's beliefs, whatever they are, were true. To believe that P is to act in ways that would tend to satisfy one's desires, whatever they are, in a world in which P (together with one's other beliefs) were true. (1984, p. 15)

Thus, one way to understand Dennett's account of mental representation is to see it as a sustained endorsement of the view that inner representations exist not as discrete entities or states, but as dispositional properties of the system. It is a philosophical account of the mind that fully embraces the tacit notion – not as a special or deviant type of representation, but as the primary notion of mental representation at the center of commonsense psychology.[3]

While no other philosopher has done as much as Dennett to develop an account of representation that accords with the tacit notion (or something very close to it), several others have generated views that can be seen as promoting the same basic outlook. Besides Clapin and Stalnaker, both of the Churchlands have endorsed and expanded upon the connectionist theme of representation within the weight matrix in their accounts of learning and conceptual change. For instance, in describing conceptual change, Paul Churchland claims:

To specify that global configuration of weights is thus to specify the global conceptual framework currently in use by the relevant individual. To change any of those weights is to change, however slightly, the conceptual framework they dictate. To trace a creature's actual path through the space of possible synaptic configurations would be to trace its conceptual history ... And to understand what factors induce changes in those weights would be to understand what drives conceptual change. (P. M. Churchland 1989, p. 232)

Insofar as our conceptual framework is a mental representation framework, Churchland is here embracing the tacit notion as the proper representational vehicle for understanding conceptual knowledge and change. A similar perspective has been pursued by Clark (1993).

Along with accounts of conceptual development, there are a variety of other areas of philosophical investigation where the notion of tacit representation is becoming increasingly significant. One such area is the field of consciousness studies. While we noted above that the notion of unconscious information processing is not to be confused with the notion of tacit

[3] Keith Frankish has pointed out that on one interpretation of his view, Dennett would actually be seen as rejecting representationalism altogether. On this reading, beliefs would not be a sort of representation-by-disposition, as I've suggested, but rather just multi-tracked dispositions. This would perhaps put Dennett even closer to Ryle and Quine than I've assumed, though it is hard to square with other things Dennett says about, say, the content and the truth and falsehood of beliefs.

representation (as we are using the term), some have tried to *explain* the former in terms of the latter. For example, O'Brien and Opie (1999) argue that a virtue of the connectionist framework is that the two types of representational structures it provides – transient patterns of activity of internal nodes on the one hand, and static connection weights on the other hand – give us a natural way to account for the difference between cognitive processing that we consciously experience and processing that we do not. Information that is only tacitly stored in the weights is not part of our phenomenal experience precisely because it is stored in this way. Nonetheless, they argue that the information is causally implicated in the processing because it is embodied by the very structures that are responsible for converting input to output. As they put it, "[t]here is a real sense in which all the information that is encoded in a network in a potentially explicit fashion is causally active whenever that network responds to an input" (1999, p. 138). For O'Brien and Opie, if we want a model of causally relevant unconscious representation, the tacit notion of representation provided by connectionist networks shows us how this is possible.[4]

One further area of philosophical theorizing where the tacit notion of representation has come to be significant is in debates about eliminative materialism. In our 1990 eliminativism paper, Stich, Garon, and myself argued that the connectionist account of information storage in the weights is incompatible with a key feature of our folk conception of stored beliefs. That key feature is their functional discreteness – folk psychology assigns to beliefs (and other propositional attitudes) a causal nature that allows them to either be active or dormant in a given episode of cognition. For example, we assume it is possible for someone to act in a manner that could have been caused by a belief we know she possesses, but in fact her action was caused by some other belief. It makes perfectly good sense to say that Mary's belief that she would inherit a large sum of money did not cause her to shoot her husband; instead it was her belief that he was going to harm her. However, on the connectionist account of stored tacit representations, this sort of causal distinctiveness is not possible for representational states. In any given episode of network processing, all of the allegedly stored information is equally involved in the processing as all of the weights are causally implicated. We argued that there is therefore nothing in the networks with which belief-type states could be identified. Thus, our conditional conclusion was that if the connectionist theory of inner

[4] Lloyd (1995) is another philosopher who tries to explain the conscious/unconsious distinction by appealing to the explicit/tacit representation distinction.

representation should prove to be correct, then an eliminativist verdict would await belief-desire psychology.

Our paper generated a large response, with some authors arguing that we had misconstrued the nature of commonsense psychology (e.g., Dennett 1991a; Heil 1991), while other authors argued we had misconstrued the nature of tacit representation (Forster and Saidel 1994). I am now inclined to agree with the latter group in one respect. In our paper we had adopted the standard line about connection weights and allowed that they were, in some sense, playing a representation role. Our point was that the weights were representations of a sort that could not serve as a reduction base for beliefs and propositional memories. I now think that in allowing that the connection weights played a representational role, we were being far too generous. In what follows, I hope to show that there is no good reason to treat connectionist weight configurations, or any other aspect of a cognitive system's functional architecture, as serving as representations in the sense suggested by the tacit notion.

5.3 A CLOSER (AND CRITICAL) LOOK

We can begin our assessment of the notion of tacit representation by asking whether or not the notion meets the job description challenge. That challenge, recall, requires an accounting of how the structure or state in question serves as a representation in (and for) the system. As with the receptor notion, we can see that there is an immediate problem. After all, it is far from clear just how the dispositional nature of a system's internal structures bestows upon them a representational function. While the actual nature of dispositions is often the subject of philosophical debate, there is no analysis that suggests that mere dispositions are, as such, a form of representation. If we say a vase is fragile, it is perhaps debatable whether we are simply making a prediction about what would happen if the vase were dropped, rather than referring to some current aspect of the vase's micro-structure (or even referring to a real property – see Mumford 1998). But no one supposes that when we assign the dispositional property of fragility to the vase, we are thereby ascribing a *representational* state to the vase. Why, then, should we suppose that by virtue of having the disposition to generate certain kinds of output, a cognitive system's functional architecture thereby embodies a type of representation? Haugeland nicely puts the problem this way: "Does it even make sense to regard the embodiments of a system's abilities or know-how as *representations*? Why not take them rather as just complex dispositional properties – acquired and subtle, perhaps – but for all that, no more representational than a reflex or an allergy?" (Haugeland 1991, p. 85)

Indeed, why not take them that way? In truth, there is not a lot in the way of detailed formal argument defending the conceptual linkage between a system's basic machinery and a type of representation. Haugeland himself attempts to forge such a connection by offering an answer to his own question that lists several reasons for ascribing a representational role to connectionist weight configurations:

> It remains the case that a network, incorporated in a real-world system (e.g., an organism), and typically encoding a considerable variety of responsive dispositions, *could have* encoded any of an enormous range of others, if only its connection weights had been different. Moreover, for that system, the actual weights consistently *determine* (fix) which abilities are actually encoded ... Finally, there are clear possibilities of malfunction or malperformance in the reliance upon and/or management of actual weight patterns ... and weight modifications in light of experience (learning) can be carried out improperly, or result in degraded performance. Thus, whether an explicit semantics is possible or not, it does seem that weight patterns can be regarded as belonging to representational schemes. (Haugeland 1991, p. 86)

Here, Haugeland is presenting a justification for treating connection weights as representational in nature that builds upon a number of common assumptions shared by many in the cognitive science community. The functional architecture of a cognitive model is viewed as representational because it is responsible for the system's unique cognitive abilities, it comes about in real systems through a learning process, and it can be the source of various sorts of malfunctioning by the system. These are features naturally ascribed to "know-how", so whatever embodies these features also embodies that know-how. And you can't get know-how without some sort of representation. We can put all of this as a more formal argument such as this:

(1) The functional architecture of a cognitive system is causally responsible for the system's cognitive abilities, it (sometimes) is acquired through a learning process, and it is often responsible for system malfunction.
(2) Therefore, the functional architecture of a cognitive system embodies a type of knowledge (often described as "procedural knowledge" or "know-how").
(3) Knowledge is impossible without some form of representation.
(4) Therefore, the functional architecture of a cognitive system embodies a type of representation.

To get the tacit aspect of representation, we only need to extend the argument by adding the observation that the functional architecture, as such, does not use any explicit representational elements:

(5) The functional architecture of a cognitive system, as such, does not explicitly represent.

(6) Therefore, the functional architecture of a cognitive system employs tacit representations.

We can call this the "Tacit Argument for Tacit Representation" because, as far as I know, few writers besides Haugeland have openly expressed it. Nevertheless, I believe something very close to this line of reasoning provides an underlying basis for a lot of talk about tacit representation. Thus, seeing what is wrong with the argument will help us see more clearly what is wrong with the tacit notion itself.

The critical inference in the argument is the move from (1) to (2) – the inference from the proposition that the functional architecture of a cognitive system possesses a number features relevant to the processing, to the proposition that the functional architecture embodies a sort of knowledge. The main problem with the argument, as I see it, is that the term "knowledge" in the second premise is ambiguous, and this leads to an equivocation. On one very weak reading of "knowledge," the move from (1) to (2) is acceptable, but on that reading the 3rd premise is false. On another, much stronger reading of "knowledge," premise (3) is perhaps correct, but on that reading the inference from (1) to (2) fails to go through. So what are these two readings of the term "knowledge"?

On the weak interpretation, the term "know" designates nothing more than the fact that a given device or system has some capacity or set of dispositional properties. This usage is on display if I were to say, for example, that my car, with an automatic transmission, knows how to shift gears on its own, or that a copier machine knows how to collate copies. When we say these sorts of things, it is readily evident that we do not mean to imply that the device or system uses inner representations or states with intentional content. We do not use "know" in this context to imply that there is some kind of knowledge base that the system somehow consults in order to perform these tasks. That is, we don't think that the causal process responsible for automatic gear shifting or collating involves the use of any sort of stored information. Instead, the ascription of "know how" in this context is just a short-hand way of saying that something is structured or designed in such a way that it has the capacity to perform some job. It is this sort of usage that Ryle (1949) focused upon in arguing that when we describe someone (or thing) as possessing know-how, we mean nothing more beyond the claim that the agent possesses certain capacities. Ryle's mistake was in thinking that *all* ascriptions of mental states are like this, which they clearly aren't. But he was certainly right that sometimes when we describe something as knowing how to do something, we are not assuming the existence of inner states with representational properties.

So, on this reading of "know", the move from (1) to (2) is trivial. (1) states that the functional architecture is causally responsible for the system's capacities (if you like, it forms the supervenience base for those capacities), and (2) simply rephrases this point by invoking the weak notion of knowledge. But then, on this weaker reading, premise (3) is false. As we just noted, no form of representation is entailed or even implied when this sort of knowledge is ascribed to a system. We don't regard automatic transmissions or collating copy machines to be representational devices (at least not with regard to these capacities). To repeat an earlier point, the ascription of mere capacities in and of itself does not entail the existence of representations. So while on this interpretation of "know", the move from (1) to (2) is warranted, the third premise is false and thus the argument is unsound.

Not so fast, you might say. While it may be true that the weaker reading of knowledge doesn't presuppose any sort of *explicit* representation, it might be argued that I haven't shown that there isn't a notion of *implicit* or *tacit* representation involved. Why not say that the automatic transmission does indeed tacitly represent knowledge of gear shifting, or that the copy machine actually does employ, in its functional architecture, some sort of tacit representation of collating information? Haven't I simply begged the question against the advocate of tacit representation by stipulating these devices don't use representations?

It is indeed true that I am assuming that for tacit representation, something more than mere dispositionality is needed. But this assumption is fully justified. To reject it would be to succumb to a general problem that I have discussed before, namely, the problem of adopting a notion of representation that has nothing to do with our ordinary understanding of representation and is utterly devoid of any real explanatory value. With this notion, representation is equivalent to mere capacity. Hence, everything with dispositions and capacities – that is, *everything* – becomes a representational system.[5] Rocks are now representational, since, after all, even a

[5] Dennett, of course, thinks that an explanatorily useful form of representationalism should not exclude complex devices like copy machines. Below, I will return to Dennett's claim that this is an explanatorily useful way of thinking about representation.

rock (in this sense) "knows how" to roll down a hill. The Representational Theory of Mind is thus reduced to the "Capacity-Based Theory of Mind" – the comically vacuous theory that cognitive capacities are brought about by an underlying architecture that gives rise to cognitive capacities! This sort of strategy offers a response to the job description challenge that, in effect, ignores the challenge altogether. Instead of an account of how something functions in a manner that is recognizably representational in nature, we are instead encouraged to think of representation as nothing more than a system's functionality. This strategy for rescuing tacit representation would come at the cost of making representation an empty and explanatorily useless posit.

Turning now to the stronger use of knowledge terminology, I think we actually do sometimes mean to imply that there is some type of distinct, causally relevant state that an agent employs as a representation in its problem-solving activities. This is the sense of "knowledge" employed when I say, after having consulted an instruction manual, "I now know how to hook-up the VCR." On this reading, encoded information is indeed represented and useable, and serves to guide my execution of a task. On this way of understanding the term "knowledge", premise (3) of the Tacit Argument comes out correct. It is impossible for there to be knowledge of this sort without the existence of representations, since without actual, useable representations of the pertinent information, there is nothing to guide the agent's activity. But now on this interpretation of knowledge, subconclusion (2) no longer follows from premise (1). As we just noted, you can't infer that there is something internal to the system that is serving a representational role (in any interesting sense), or that encoded information is exploited by the system from the mere fact that the system possesses capacities that are due to an underlying functional architecture. Moreover, the various aspects of the functional architecture appealed to by Haugeland and mentioned in the first premise fail to imply that a stronger sense of knowledge must be at work.

Consider Haugeland's point that the architecture, "encoding a considerable variety of responsive dispositions, *could have* encoded any of an enormous range of others . . ." This certainly seems correct, but there is no reason to suppose this alone gives rise to internal representations. The functional architecture of *any* system could embody a range of other functional dispositions if the functional architecture had been different. That's the way functional architectures and dispositional properties work. Yet these counterfactual conditions don't lend any support to the proposition that the system in question is thereby representational in nature. Had my automobile's transmission been different, it certainly would have embodied a different range of responsive dispositions. But I see no reason to take that as evidence that the transmission is in some way representational or employs tacit representations.

Or take the point that the underlying architecture is sometimes the source of system malfunction. Again, this is true, but how does this warrant any type of representational analysis of the architecture? The inner design of automatic transmissions and copy machines can also break down and give rise to malfunctioning systems. But this alone doesn't support treating these things as employing representations. Of course, if the malfunctioning involved some form of misrepresentation, then *that* type of malfunction would presuppose the existence of representations. However, this would require that we first establish the existence of representations. As we saw in the last chapter, you can't establish the presence of misrepresentation by simply establishing the presence of system malfunction.

But what about the point that the functional architecture is sometimes the result of a process that is normally characterized as "learning"? We saw earlier that connectionist weight configurations in feed-forward nets often develop through the employment of a "learning algorithm" like back-propagation. During this process, repeated trials gradually adjust connection weights through an error signal that is used to modify individual connections in proportion to their contribution to faulty output. Through this sort of process, connectionism is traditionally thought to offer a biologically plausible account of at least one sort of cognitive learning. But if learning is actually going on, mustn't it result in the acquisition of knowledge, information, and representational states of some form?

In thinking about this point, it is first important to recognize that we should not feel obligated to adopt a particular characterization of a cognitive model or theory just because it is described that way by the model's authors. Indeed, a central theme of this book is that researchers and cognitive modelers, along with philosophers, not only can but sometimes do misdescribe what is going on in a proposed model of the mind. This danger is especially real with intentional characterizations of mechanisms and processes. Consequently, the mere fact that connectionist researchers often describe weight adjustments as "learning," does not, by itself, establish that we should think of them this way. What is clear is that back-propagation and other such algorithms systematically and purposefully produce weight adjustments that eventuate in networks capable of performing impressive input–output transformations. Yet systematic and purposeful modification of a system that eventuates in something desirable doesn't, as such, require a process properly described as learning. Jogging and exercise will systematically modify my lungs and muscles in a way that is both deliberate and desirable. But these modifications don't really qualify as a type of learning. Of course, my lungs and muscles don't comprise cognitive systems. But I see no reason to suppose that every sort of positive modification of a cognitive system – including my own mind – should be viewed as a form of learning.

Yet even if the weight modifications are regarded as a type of learning, this still shouldn't lead us to conclude that the modifications bring about a form of internal representation. It seems possible for there to be various developmental processes that one could describe as learning but that don't involve the acquisition of any new states playing a representational role. Behaviorist accounts of learning are, after all, accounts of *learning*, even though they typically reject, quite openly, the idea that the process results in mental representations. Consider a process that modifies an information processing system not by adding new structures or internal elements, but by simply speeding up or increasing the efficiency of existing mechanisms and processes. This might be the sort of learning that underlies the development of a skill or enhanced facility in performing some athletic task. While such a process might qualify as learning, nothing changes except the rate, direction, and smoothness of the processing. In fact, connectionist weight modifications resemble this sort of process. Weight tuning looks more like the sort of transformation that simply improves non-representational processing, and less like a process that somehow develops and stores tacit representations. The critical point is that the mere existence of a learning process needn't force us to conclude that the system acquires new information or representational elements, even in a tacit form.

Thus, the popular descriptions of network learning don't provide a strong motivation for treating the functional architecture of a system as embodying tacit representations. Since none of the other features mentioned by Haugeland support a representational characterization of the functional architecture either, the first premise of the Tacit Argument, as it stands, fails to support the proposition that there are tacit representations in cognitive systems. Yet perhaps there are other considerations, apart from those mentioned by Haugeland, that can take us from the dispositional nature of a system's internal machinery to the conclusion that this machinery embodies tacit representations. For example, what about Dennett's claim that ascribing tacit representations to computational systems is "very useful and predictive"? If you are playing against Dennett's chess-playing system, it does indeed seem that it would be beneficial to treat it as thinking it should get the queen out early. I've been suggesting that the tacit notion of representation is explanatorily and predictively unhelpful. Yet Dennett's case seems to demonstrate just the opposite.

Dennett is correct that the sort of characterization of a computational system he mentions can be quite useful and predictively valuable. But as far as I can see, this has nothing to do with the ascription of representations. When the programmer says "It thinks it should get the queen out early," the predictive value of such a characterization does not extend beyond the predictive value of assertions like, "It employs a strategy of early queen deployment" or "The system is designed so that when it plays, the queen typically comes out early." What makes the programmer's characterization useful is not that it is intentional or representational in nature, but that it is *dispositional* in nature. The programmer is implying that the system has a certain dispositional state, and invoking dispositional states is indeed predictively useful. The critical point is that the concept that is doing all of the relevant predictive work is the concept of a disposition, not the concept of a representation, or even some hybrid concept like "representation-by-disposition." Now we can, of course, always frame talk about dispositions in representational language. For example, we can describe the vase's fragility by saying that the vase thinks it should break when dropped, or perhaps by saying that it tacitly encodes the rule 'IF DROPPED, THEN BREAK'. But there is no predictive value in adopting this sort of language beyond the predictive value provided by saying the vase is fragile. In this and many other cases where tacit representations are assumed to play a pivotal theoretical role, what is actually playing that role is nothing beyond an ascribed dispositional state.

A similar skeptical point has been made by Cummins (1986), who assesses Dennett's remarks in light of the distinction between rules that are actually represented by a system, and rules that are not represented but instead executed. Cummins notes that the attempt to identify representational states with mere dispositions leads to an utterly bogus notion of representation. Calling an imaginary computational system like the one Dennett describes as "CHESS," Cummins notes:

> But CHESS ... simply executes the rule without representing it at all, except in the degenerate sense in which rubber bands represent the rule IF PULLED THEN STRETCH and masses represent the rule COALESCE WITH OTHER MASSES. Like the rubber band, CHESS simply executes its rule, and executing that rule amounts to having a behavioral disposition to deploy the queen early ... CHESS (as we are now imagining it) does not represent the rule requiring early queen deployment,

nor is anything with a comparable content available for reasoning or epistemic assessment. (1986, p. 122)

Cummins extends this point to make a general point about the identification of tacit representation with procedural knowledge in classical systems:

A frequent reply to this point is that our system has *procedural knowledge* to the effect that the queen should be deployed early in virtue of executing a program containing the rule . . . I think talk of procedural knowledge has its place – it's the case . . . in which [it is] explicitly tokened in memory – but a system's procedural knowledge is not knowledge of the rules it executes. The rules a system executes – the ones making up its program – are not available for reasoning or evidential assessment for the simple reason that they are not represented to the system at all. (1986, p. 122–123)

Unlike myself, Cummins is more inclined to suppose that, even though there are no tacit representations, there is nevertheless an important notion of inexplicit information to be found in the functional architecture of such systems.[6] But Cummins is correct that the computational system Dennett describes is properly viewed as possessing dispositional states, not tacit representations, and it is really these dispositional states that are invoked in useful explanations and predictions.

Yet what about Ryle's regress challenge? Recall that both Dennett and Clapin justify the invoking of tacit representations by arguing that they provide a means of avoiding the sort of regress that Ryle claimed threatens any representational theory of the mind. If Ryle is right that the employment of explicit internal representations requires yet deeper representational states, then there must be some strategy for stopping the possible regress. Dennett and Clapin claim the way to do this is to appeal to tacit representations because, apparently, the use of tacit representations does not require further sophisticated operations or processes. So perhaps this is a strong motivation for thinking that the functional architecture of any explicit representation-using system (such as those described by the CCTC) also embodies tacit representational states.

However, there are two serious problems with this argument. First, there is no reason to think you can break a Ryle-type regress by appealing to tacit

[6] Cummins puts matters this way: "Nevertheless, and here is the main point at least, even though the system doesn't represent such rules, the fact that it executes them amounts to the presence in the system of some propositionally formulatable information, information that is not explicitly represented but is inexplicit in the system in virtue of the physical structure upon which program execution supervenes" (p. 124). Presumably, Cummins would have to admit that a rubber band also carries this sort of unrepresented but "propositionally formulatable information," executed by stretching when pulled.

representations. The reason is that, *qua* representations, tacit representations still need to play some sort of representational role. If Ryle is right, and it is indeed impossible for something to play a representational role without deeper mentalistic (i.e., representational) processes, then tacit representations would require yet deeper representations every bit as much as he claims explicit representations do. Indeed, as some have argued, it would seem that the more tacit or implicit the representation, the *more* sophisticated or mentalistic the internal sub-system would need to be to extract the relevant information (Kirsch 1990). Going the other way, if it is possible for the functional architecture to represent tacitly without still more representations, then it is unclear why it wouldn't also be possible for structures to represent *explicitly* without still more representations. To break the regress, we would need to show not that the background functional architecture embodies tacit representations, but that it embodies *no* representational states whatsoever. So Dennett and Claplin's solution to the regress problem is not much of a solution at all.

Secondly, an alternative solution to the regress problem is readily available and undermines the need for tacit representation. Contemporary cognitive models reveal how Ryle was just plain wrong to think that the employment of internal representations demands yet deeper representational states. It is possible to avoid Ryle's regress by showing how a surrounding functional architecture can employ structures that serve as representations without the architecture *itself* being representational in nature. There are different ways this might go. For example, in chapter 3 we caught a glimpse of what it might mean to have a mindless computational architecture that is, despite its own level of ignorance, nevertheless using a model of some domain, and thus using S-representations. The relevant computational architecture is perhaps complex and elaborate. But being complex and elaborate is not the same thing as embodying representations. Sophisticated operations can be carried out by inner subsystems that execute surrogative reasoning (and therefore makes use of representational states) even though the sub-systems are themselves utterly void of their own inner representational states. The central point is that components and elements of a cognitive system that make the use of a map or model possible don't themselves require their own inner intentional states. In chapter 6, we'll see how this might go for the mechanical use of a map for navigation purposes. The bottom line is that Ryle's regress worry is only a pseudo-problem.

So far, my argument has been to challenge the Tacit Argument for Tacit Representation to show that there is no compelling reason to buy into tacit representation. My claim has been that the argument is unsound because of the equivocation on the term "knowledge," and because the popular reasons for treating the functional architecture as embodying a type of tacit representation are no good. Yet it would not be difficult to reconstitute many of these points into a more direct and stronger argument for the position that so-called tacit representations are really not representations at all. The more direct argument would go something like this:

(1) When investigators invoke tacit representations in explaining cognitive processes, they are (at best) referring to conventional dispositional properties of the system's functional architecture.
(2) Dispositional properties, as such, aren't representations.
(3) Therefore, so-called "tacit representations" aren't really representations at all.

I am assuming that the second premise is fairly uncontroversial, as demonstrated by the absurdity of treating anything with a disposition as a representational system. As we saw above, it is clearly silly to treat a vase's fragility as serving to represent something like the conditional "if dropped, then break". Thus, since the argument is valid, the premise doing all of the work is the first one. In the case of receptor representation, we saw that structures or states described as fulfilling a representational role were in fact actually serving as something more like relay circuits or simple causal mediators. We can now say something similar about tacit representations. When tacit representations are invoked, all of the legitimate explanatory pay-off stems from a perhaps inadvertent appeal to the dispositions embodied in the functional architecture of the cognitive system. Just as the role assigned to receptor representations is actually more like the role associated with some type of causal relay, so too, the role tacit representations play in cognitive processes is nothing beyond the sorts of roles that can be assigned to dispositional properties. We have already seen how this point applies to classical systems, as illustrated by our analysis of Dennett's claims about the chess-playing system. A similar point can be made about the tacit representations alleged to exist in the weights of connectionist networks. For example, if it is claimed that a network can perform a discrimination task because it has acquired a tacit encoding of conceptual knowledge in the weights, the theoretical value of such a claim is entirely derived from the weight assignments having acquired dispositions to generate one type of response when presented with one sort of input, and a very different type of response when presented with input of a different sort. There is no additional theoretical pay-off in pretending the weights play some further (largely unspecified) representational role that goes beyond the acquisition of these propensities.

Or when Rumelhart and McClelland (1986b) present a connectionist model devoted to learning past tense verbs and suggest its weights have acquired tacit representations of linguistic rules, what we see, on closer inspection, is just a network whose weights produce a response profile that mimics the stages children go through in learning regular and irregular forms. Yet we don't see any explanatory need for the supposition that there are also representational states hidden in the weight matrix and playing some additional role in the processing. There is no more reason for thinking the weights play a representational role than there is for thinking that the micro-structure of a fragile vase or the molecular constitution of a rubber band play representational roles.

One way to explore this point further is to take a harder look at the notion of superpositionality that has become so central to understanding tacit representation. It is far from a given that the idea of superposed representations is even fully intelligible. Just how it is that truly distinct, specifiable items of content can be simultaneously represented by the *exact same* physical and computational conditions is not at all clear. By contrast, the idea of an architecture acquiring superposed dispositions is not at all problematic. If superpositionality is a real feature of connectionist models, we need to reconsider just what it is that is being actually superposed.

First, consider the questionable intelligibility of the concept of superpositional representation. Various promising strategies for explaining how distinct intentional objects can be simultaneously represented in the same structures all wind up abandoning the basic idea behind superpositional representation. For example, one might suppose that what is represented by the static weights is a single complex sort of content. We might suppose that rather than representing a set of propositions or distinct conceptual elements, the weights actually encode a single, complex conjunctive (or disjunctive) proposition, or a single super-concept. But this is clearly not what people have in mind by superpositional representation. Instead, distinct and separate intentional objects are thought to be encoded by a single global state. Another possibility might be that the weights function as a complex state with identifiable features that individually stand for particular things or propositions. Haybron (2000), for example, illustrates this form of representation by appealing the nature of a complex sound wave. Individual sound waves, such as one resonating at 1 kHz and another resonating at 2.3 kHz, could encode distinct propositions and then combine to form in a single complex wave. But, as Haybron himself notes, this is a situation where distinct chunks of information are assigned to specific parts of a more complex phenomenon. If distinct, identifiable properties of the connection weights somehow represented specific items, then this might give us an intelligible account of how the weights could represent separate, distinct intentional objects. But this is not the superpositional story. With superpositional tacit representation, individual contents don't correspond with *any* distinct aspect of the connections or their weights. If they did, this would actually provide us with a complex form of explicit representation.

Haybron (2000) discusses another scenario that he claims provides a better example of how superpositional representations can occur and do real explanatory work. While the case is complicated, the basic idea is fairly easy to grasp. We can imagine a conventional computer that stores items of information in a way that requires the system to look up other stored items of information while performing some computation. For example, suppose that to decipher someone's age, the system must carry out a calculation that involves checking values in specific registers; and to determine another person's salary, the system must again check the exact same values in the exact same registers. Haybron's claim is that in such a case, information about both the first person's age and the second person's wage would be superpositionally encoded by the values in the same registers. He suggests that this reveals how even classical computational devices can make use of information that is encoded in a superpositional manner.

Haybron's example is related to an account of tacit representation that has been offered by David Kirsh (1990). Kirsh suggests that we can determine just how implicit a representation actually is by attending to the amount of computational resources and energy required to retrieve the encoded information. If the system can retrieve the content easily, then the representation is explicit. If not, then the more difficult the information retrieval, the more implicit the representation. For example, Kirsh notes that for us the numeral "5" transparently designates the number 5 and thus explicitly represents it. But the function $5\sqrt{3125}$ requires work to derive the number 5. Hence, it only implicitly represents 5. Kirsh's distinction might be thought to provide a means for making sense of superpositional encoding that is similar to Haybron's suggestion. On this way of understanding representation, one could say that $5\sqrt{3125}$ not only implicitly represents the number 5, but it also superpositionally represents a variety of other things, such as the function $2+3$, or the number of fingers on my left hand.

However, this way of thinking about the representation of information stems from a confusion about the nature of the intentional relations. The number 5 is not *represented* by $5\sqrt{3125}$; rather 5 is the solution for $5\sqrt{3125}$. The function $5\sqrt{3125}$ no more represents the number 5 than the number 5 represents $5\sqrt{3125}$, and/or $3+2$, and/or the question, "how many fingers are on my left hand?" The relation between a representation and its intentional object is not the same as the relation between a function and its solution, or, for that matter, between a statement and whatever the statement entails. Of course, one could certainly adopt a convention that treats the symbols that represent a function as also representing the function's solution. But that would make the representational relation derived from convention, not from these mathematical or entailment relations. So what Kirsch is describing is not a distinction between explicit and implicit representation; instead what he is describing is a distinction between representations on the one hand (e.g., the numeral "5"), and functions that have derivations on the other. Generating those derivations can indeed involve varying degrees of computational resources, but that is not the same thing as extracting encoded information.

Returning to Haybron's example, a computational system can certainly employ the same values in different types of calculations and computations, but those calculations and computations are not represented by those values. It is clearly one thing for the same explicitly stored bits of information to play various roles in different computations or derivations, it is quite a different thing to claim that those values serve to simultaneously represent a collection of different contents. The numeral "5" represents the number 5, which is also the value of $5\sqrt{3125}$, the number of fingers on my left hand, the number of days in a typical work week, and so on. The numeral "5" does *not* superpositionally represent all these other things – it just explicitly represents 5.

A similar point applies to suggestions that tacit superposed representations can be found and recovered in activation patterns through some form of tensor product analysis. Smolensky (1991) has suggested that representations with constituent structure can be generated through a process of tensor addition and multiplication. One set of nodes might represent syntactic role, such as noun, and another set of nodes could represent a filler of this role, like "John." The values of the activation patterns of these nodes might then be multiplied and the product can be represented in a set of "binding nodes." From this pattern of activation, it is possible to reverse the process and "recover" what is said to be tacit, superposed representations of the role and filler. If we know the value of the explicit tensor product representation, and the value of the explicit role representation, then, of course, through a process of tensor division, we can determine the value of the explicit filler representation. But this doesn't mean that activation patterns of units corresponding to tensor products tacitly *represent* their multiplicands, any more than normal numbers tacitly represent their multiplicands. Instead what it means is that the values of these tensor product representations can serve as input to a mathematical operation – in this case, tensor division – and that operation can then serve to generate an explicit representation of either the role or filler. The value, like any number, certainly has a huge range of dispositional properties, corresponding to all of the different mathematical operations or mapping functions in which it can play a role. But these dispositions are not a form of *representation* embodied within the value. In fact, if things tacitly represented whatever they have a disposition to generate via some mapping function, then given that anything can be mapped to anything else, everything would tacitly represent everything else!

One further way someone might try to explain superpositional representation would be to point out that a single set of symbols can often have many different meanings according to different interpretation schemes. After all, words and even sentences can be fully ambiguous, and people can send encrypted messages that are hidden in a single text that appears to be about something entirely different. If fixed linguistic structures can have different contents, why not a single weight configuration? The answer is that in the linguistic case, there are alternative decoding or translation processes that are always employed to extract the different meanings from the same set of symbols. Spies, for example, must use some alternative interpretation scheme to extract hidden messages from texts that have conventional interpretations. But in typical connectionist processing, there are no alternative decoding operations that serve up different interpretations for the weights. Indeed, there are no translation conditions or interpretation operations in the processing at all. There are no procedures that play the role of information retrieval, or data recovery, or knowledge look-up, or any similar sort of process. All we have is a wave of spreading activation, passing through a fixed set of weighted connections. Even if we granted that there was some sense in which different contents were represented by the same set of connections, there doesn't appear to be anything that distills the relevant representational contents during the processing.

In response to this last point, it is sometimes suggested that the distinctive responses of the network generated by different inputs are *themselves* a sort of information extraction process that finds and uses the right content from the network's static weights. It is in this way that distinct superposed contents can become causally implicated and the network is able to provide the right responses to certain inputs. An advocate of such a view might suppose that I am being too traditional, or perhaps insufficiently imaginative, in my understanding of how stored information can be brought to bear on the processing. Isn't there at least some sense in which a network's operations are guided by tacitly stored chunks of information that become activated by the relevant inputs to the network?

Sometimes in science and philosophy, real insight occurs when we see how something can happen in a radically different way than previously imagined. However, sometimes real confusion can occur when we try to force a pre-conceived image onto processes that are best understood by abandoning that image. I believe that the latter (and not the former) describes attempts to make sense of network activation waves as a type of knowledge recovery process. So, to answer the question at the end of the last paragraph, no, actually, I think there is *no* helpful sense in which the wave of activation can be viewed as an information-extraction or look-up process. The reason is fairly simple: the mere passing along of a simple activation signal from one unit to another is not, as such, a process that we can seriously regard, at any level of analysis, as an information retrieval or representation interpretation process. If node A excites node B to a degree of 75 percent because that is the weight of the connection between the two units, then there is really no point in describing the weighted link as encoding "superposed information" and the activation as "retrieval of that information." Unlike the situation with various CCTC models, where there is an actual causal/physical process involving the retrieval of structures from a register that have the function of representing, here we have a process more akin to the switching on of a light. Now of course, there are fanciful descriptions that we can apply to any causal sequence. We can certainly call the activating of units "information retrieval," just as we can call it "deductive inference," or "guessing," or "conceptualization." The turning on of a light switch can be given such characterizations as well ("tacitly encoded in the electrical wiring is the knowledge that lighting is needed, and this information is retrieved every time the switch is turned on"). But all of these are unhelpful adoptions of the intentional stance toward more basic processes – something that, as we noted in chapter 1, is always possible but not always explanatorily justified. It is simply explanatorily gratuitous and metaphysically unwarranted to treat the functional role of the weights as one of storing and facilitating tacit superposed representations.[7]

[7] Consider the long-standing and useful distinction between processes that exploit and are guided by stored information, and processes that are instead "hard-wired." To qualify as the former, there needs to be something remotely resembling stored representations, and some sort of process that counts as the retrieval and use of those representations. In many classical computational systems, this comes about by processes that involve structures or symbols that are literally retrieved from memory registers. With other architectures, it is perhaps less clear whether the processes qualify as using encoded representations. But it doesn't follow from these grey cases that anything goes, or that we can't nonetheless accurately describe some systems as *not* using stored data; that is, as merely hard-wired. A hard-wired system would be one that simply carries out operations in a manner that lacks operations recognizable as information retrieval. Connectionist feed-forward processing is a paradigmatic example of processing of this nature.

These criticisms will no doubt fail to convince many die-hard connectionist representationalists, who might be inclined to view me as simply begging the question against what they see as an unconventional form of information retrieval. Perhaps, then, the best way to make my point is to make an appeal in burden court. When a radical new way of conceiving of a certain process is proposed, then surely the burden is on those endorsing the new outlook to show that it is legitimate and intelligible. Thus, for those who claim that a spreading wave of activation involves the retrieval and employment of superposed representations or superposed representational contents, the burden is upon them to make this intelligible – to explain how this works. In fact, they must explain this in a way that does not destroy a number of useful distinctions (such as the distinction between representational systems and non-representational systems) or that doesn't render a notion of representation that is explanatorily useless and alien to our ordinary understanding of what a representation is. As far as I can see, this burden has not been met, and it is difficult to see how it *could* be met.

I have dwelt upon superpositional representation because it has become so integral to the way people think about tacit representation. As we've seen, it is far from clear that we can even make sense of the idea that structures like connection weights serve to simultaneously represent different intentional objects. By contrast, the idea that a range of distinct dispositional properties can be instantiated by a single physical system or entity is not problematic at all. When we ascribe a specific dispositional property to something, we are claiming that its nature is such that if certain conditions obtain, it will react or behave in a certain way. Since different conditions can give rise to different reactions, a single underlying architecture can embody different dispositional properties. So, for example, the molecular lattice structure of a glass is responsible for the glass's breaking if dropped and also for its containment of liquids. Thus, the same lattice structure embodies (or underlies) the dispositional properties of fragility and impermeability. We can see how the same basic physical arrangement can simultaneously embody a number of different dispositional states. Hence, there is no problem in saying that a single weight configuration of a connectionist network embodies a range of different dispositions, since different inputs to the network will generate different outputs. In other words, different dispositions can be superposed onto the same single weight configuration. Suppose we train a network to give an affirmative response when presented with a representation of "cats have fur" and a negative response when presented with "dogs have scales" (Ramsey, Stich, and Garon 1990). While it is utterly mysterious how the single weight configuration can simultaneously represent each of these distinct propositions, there is no mystery at all in seeing how a network can acquire a weight configuration that has two dispositional properties that are manifested by these two different responses. This suggests that superpositionality is a property not of representation, but of dispositional properties.

Let's briefly reconsider Kirsh's account of implicit representation. Recall that Kirsh appears to confuse the representation of some value with the *derivation* of that value when he suggests equations represent their solutions. But Kirsh is right to note that different derivations require different degrees of computational resources and exertion. One way to understand Kirsh's distinction is not as a distinction that corresponds to degrees of implicitness of representation, but as a continuum describing the degrees of difficulty associated with the generation of explicit representations. If the system is simply asked to produce a representation of 5, and it possesses a symbol that explicitly stands for 5, then the task is trivial. If, on the other hand, the system is asked to find the cubed root of 125, then it will need to do some calculations to generate the same symbol. In the second case, 5 is not implicitly represented in the calculations or even in the system's capacities. Rather the explicit representation of 5 (say, the symbol "5"), is generated by the system's computational capacities. Sometimes the system is up to the task, and sometimes it isn't. But the key point is that the state of affairs Kirsh identifies is not a representational state of affairs, but a dispositional state of affairs. What he calls the "implicitness" of a representation is actually the degree of complexity underlying the system's capacities to generate actual (explicit) representations.

Recall that at the outset of this chapter we identified a commonsense notion of implicit knowledge whereby people are said to know things they have never thought about, but would immediately assent to if given the opportunity. As we noted, one natural way to interpret this sort of ascription is to view it not as the assignment of a current but implicit sort of representational state, but rather as the assignment of a disposition to acquire an explicit representational state if certain conditions are met. If someone said that Joe knows that hang-gliders cannot go into outer space, she might mean only that Joe's cognitive machinery is such that he would immediately acquire a belief with the content "hang-gliders cannot go into outer space" if prompted in the right way. What I would now like to suggest is that this analysis should be extended more generally, to the notion of tacit representation that has been the focus of this chapter. When we say that someone tacitly knows how to ride a bike, or that a connectionist weight configuration tacitly represents a certain rule, we should interpret this sort of talk as an oblique reference not to representational states, but to the sorts of dispositional properties of cognitive systems. In other words, there are really only two ways to make talk about tacit cognitive representations meaningful. One is to suppose that there is indeed some sort of explicitly represented information inside of the cognitive system, yet its employment is perhaps unconscious, in some sort of dormant state or perhaps in some sense inaccessible. The other is to suppose that we aren't really talking about representations at all, but instead referring to dispositional states in a manner that is misleading. What I am denying, however, is that there is an intelligible third possibility – that the dispositional nature of the cognitive architecture is itself playing a representational role. Thus, when cognitive scientists claim the brain uses these sorts of states, they are making a fundamental error in their accounting of cognitive processes.

It is worth noting that my critique of the notion of tacit representation is in many respects similar to attacks on the idea that genes serve as information encoding structures. It is commonplace to hear prominent biologists, such as Maynard-Smith (2000), suggest that DNA sequences should be seen as encoding instructions for the development of various traits, and that they are therefore best viewed as an information storage device not unlike what is found in computational systems. Some, such as Paul Griffiths (2001), have challenged this perspective, arguing instead that information talk is misleading, and that genes are best viewed as non-intentional causal contributors to the development of phenotypes. One type of argument supporting skepticism about genetic information mirrors in certain respects our earlier point that equating representations with dispositions would force us to call virtually everything a representational system. For example, Griffiths notes that because there are many causal factors (besides inherited genetic material) that contribute to the development of specific traits, we would have no principled reason to deny these other factors the role of information-encoders too. This leads to a sort of reductio, as it suggests that methylation patterns or incubation temperatures are also information-exploiting mechanisms. Griffiths concludes that there really is a genetic code in the sense of specific sequences of amino acids, but that, "'information talk' in biology has nothing to do with genetic code . . . There are numerous important differences between what DNA does in development and the roles played by other causal factors, but these differences do not map onto a distinction between information and material causation. The present atmosphere, in which information talk is only applied to genes, makes that way of talking highly misleading" (2001, pp. 409–410). If in genetics, "information talk" is nothing more than a way of talking about causation, then, as Griffiths notes, we can be misled into thinking that genes are also in some sense serving as representational structures. My claim is that a similar mistake has occurred in cognitive science, where "information talk" is applied to the functional architecture of cognitive systems, *just because* the architecture has certain dispositional properties.

5.4 CONCLUDING COMMENTS

In an earlier work attacking connectionist notions of representation (Ramsey 1997), I critically assessed the notion of tacit representation in connectionist weights by asking three questions: (1) are the weights regarded as representations for reasons that justify talk of representation in other models or, alternatively, for reasons that are unique to the style of explanation offered by connectionism?; (2) is the notion of representation strong enough to be informative?; and (3) would anything be lost if we stopped regarding the connection weights as representations altogether. Regarding all three questions, I argued the proper answer is "no."

In this chapter, I've extended much of this assessment to the general notion of tacit representation that appears in various cognitive theories. Thus, the reasons that motivate us to call structures representations in other systems, as we discussed in chapter 3, can't apply to tacit representations since for something to serve as representational input or output to a system, or as elements of a model, it needs to function as explicit representations. Moreover, it does not seem that there is anything unique to the theoretical commitments of cognitive theories invoking tacit representation that actually requires us to view the functional architecture in this way. Indeed, as we have seen, the motivation appears to stem from the largely philosophical and problematic assumption that it is reasonable to view dispositional properties of a system's functional architecture as representations. Second, as we've already noted, this notion of representation is absurdly weak. Since just about any system has dispositions, or some sort of inner functional architecture, then just about any system (like automatic transmissions and rubber bands) would employ representations in this sense. Finally, nothing of explanatory significance would be lost if we dropped the notion of tacit representation altogether. Doing so would not prevent us from claiming that, say, weighted connections are modified during a developmental phase, or that the functional architecture of a system gives rise to its computational capacities in various ways, or that a given system has a wide range of dispositions, and so on. The invoking of tacit representations is, in a sense, a theoretical add-on that is not an essential aspect of most cognitive models. If there were a story about how the functional architecture of a system actually *functions as* some form of representation (without the existence of any explicit representations) then perhaps a notion of tacit representation would be valuable. But as far as I am aware, there is no such story, at any level of analysis.

The proper verdict for the concept of tacit representation, then, is similar to the one we saw with the receptor notion. Notwithstanding its growing popularity, the tacit notion is a fundamentally flawed theoretical posit of cognitive theories. When it is invoked, the structures picked out to serve as representation of this sort – things like weighted connections in a network – are not up to the task. In fact, it is doubtful that anything could be up to the task because it is doubtful that anything could actually function as a representation in this tacit sense. Instead, it appears that theorists are confusing the possession of a dispositional state with the possession of a representational state, and this confusion gives rise to the mistaken characterization of a proposed model's inner workings.

None of this is meant to suggests that the models or theories can't be eventually reworked so that these conceptual mistakes are corrected. Moreover, it might turn out that the underlying architecture of cognition is indeed just like these theories claim, apart from the confused appeal to tacit representations. What it does suggest is that there might be a lot less representing going on in the brain than what is currently assumed. If it should turn out that the connectionist framework presents an accurate picture of how the brain acquires and manifests various cognitive capacities, then those capacities are not driven by representational structures. What might such a scenario mean for the future of cognitive science, the cognitive revolution and, ultimately, our basic understanding of ourselves? In the next chapter, we will look at some possible answers to these questions.

6

Where is the representational paradigm headed?

In this chapter, I want to do three things. First, in the next section, I want to make another pass at showing that when we think carefully about what it means for something to function as a representation in a cognitive system, we find that some notions measure up and others clearly do not. This time my analysis will be from a different angle. Instead of looking at representational notions as they are employed in different cognitive theories, I want to offer a more direct comparison between different types of representation that are functioning inside of the same simple system. My hope is that this will make clearer just how and why certain notions belong in scientific theories while the others do not. Second, while cognitive science is best described as having embraced a representational paradigm, there is a line of research that is moving in a different direction and that has generated considerable controversy. Dynamical systems theory is alleged to offer a framework for understanding cognitive processes that differs greatly from both the classical computational tradition and connectionism. One of the key differences emphasized by many dynamicists is their rejection of representation as an explanatory posit. This claim, in turn, has generated a wave of pro-representational advocacy. In section 6.2, I will look at how these debates bear on my own account and challenge some of the ways representationalism has been defended by its proponents. Finally, I want to explore a few of the ramifications of my analysis for our understanding of the mind. If my analysis is right, then cognitive science is secretly (and non-deliberately) moving in a direction that is abandoning representationalism. Does this mean we are returning to a behaviorist, or neo-behaviorist science of the mind? And what does this suggest about the status and prospects of folk psychology? If the best theories of the mind don't posit states that function as representations, what does this suggest about belief-desire psychology? In section 6.3, I will offer some tentative answers to these and other questions.

6.1 RECEPTOR AND S-REPRESENTATION REVISITED

In chapters 3 and 4, we looked at what are arguably the two most significant and important ways to think about inner representation in cognitive science – the notion of S-representation and the notion of receptor representation. In these chapters, the analysis focused on the sort of physical conditions and relations that have been assumed to bestow upon an internal state the status of representation. As we saw, the conditions that underlie S-representation successfully answer the job description challenge (which in turn suggests this notion is scientifically respectable) while the conditions that typically underlie receptor notion do not (suggesting it is not a valuable scientific posit). To some degree, this reverses contemporary conventional wisdom on the matter. Among many philosophers, but also among many cognitive scientists, the popular trend is to try to understand representation in cognitive systems by focusing upon the way an internal state co-varies with external conditions. Model-based or isomorphism-based conceptions of representation are certainly out there, but they are often viewed as less promising than nomic-dependency accounts. This might lead some to conclude that my analysis must be flawed or unfair to the receptor account. Thus, it would be helpful if we could make a more direct comparison between the two notions, especially if such a comparison could shed a little more light on just why one notion works while the other one doesn't. In this section, my aim is to offer just such a comparison.

In our earlier discussion, it was noted that these two representational notions have a number of non-mental analogues which have, no doubt, provided a basis for thinking about mental representation. Philosophical and empirical perspectives on cognitive representation often begin with our understanding of non-mental representations – the sort of representations that we encounter and employ "out there" in the world. So conceived, the project is simply this: look at the way representation works in the physical world and then use what you learn to fashion a naturalistic account of mental representation. I have no deep argument for the claim that our popular notions of cognitive representation are based upon notions of non-mental representation, but it strikes me as an eminently plausible assumption to start with. With the receptor notion of representation, it seems reasonable to suppose that its appeal is largely due to our regular use of external natural indicators to find out things about the world. We exploit these indicators sometimes in their natural state, as when we see smoke and infer the existence of a fire, and sometimes they are harnessed in things like gauges and monitoring devices, as with the mercury column or bi-metallic strip in a thermometer. It is presumably our familiarity with these external indicator-type representations that provides a motivation for thinking something similar is at work inside the mind/brain. The same can be said about the S-representation notion. No doubt its appearance in theories of cognition is at least partly based upon our use of external physical maps, models, or simulations. So a useful way to regard the invoking of representational states in cognitive theories, and also the philosophical project of explaining cognitive representation in naturalistic terms, is to see all of this as using these familiar examples of non-mental representation as prototypes for understanding and explicating mental representation. It is this enterprise that I would like to explore in greater depth and use as a backdrop for our comparison of the two notions.

As I emphasized in chapter 1, there is a problem with appealing to non-mental representations to explain and understand mental representation. The problem is that to play a representational role, non-mental structures apparently require a full-blown cognitive agent that uses them in a certain way. While it is true that many non-mental representations are grounded in purely natural conditions that serve to forge a connection between the representation and its intentional object, the connection alone is not sufficient to make something into a representation. For example, the nomic dependency relation between smoke and fire is not enough to make a column of smoke into a representation of fire. For the smoke to play this role, it needs to be read or interpreted as implying fire by cognitive agents like ourselves. Similarly, the mere fact that squiggly lines on a sheet of paper are isomorphic to some path is not enough, presumably, to make the lines represent the path. These lines acquire representational status only after we (or some similarly sophisticated cognitive systems) use them to find out about some aspect of the world. So, apparently, the column of smoke and the lines on the paper come to serve as representations only once minds like ours employ them to gather information – to find out about the presence of fire or the nature of some terrain. But, recall, a sophisticated mind is impermissible in any account of cognitive representation. Any explanation of the mind that appeals to little inner minds as inner representation users runs the risk generating a regress of inner representations. Hence, any account of mental representation built from the way we use non-mental representations has to find a way to deal with this regress worry.

In philosophical work on mental representation, there have been three different proposals for handling this problem. Two of these are suggestions from Dennett that we have already looked at. The first strategy, which we discussed in the last chapter, is to make an appeal to allegedly tacit representations stored in the system's functional architecture. These are then thought to allow for the use of explicit representations without the need for a sophisticated, regress-producing homunculi (Dennett 1982; Caplin 2002). As we saw, this strategy doesn't succeed because a) you can't avoid a regress by simply making representations tacit or inexplicit, and b) there are no such things as tacit representations. The second strategy, also proposed by Dennett (1978), is to accept the need for inner homunculi that use representations, but avoid a regress by decomposing the interpreting homunculi into increasingly less sophisticated sub-systems. Representation-using homunculi can be explanatorily harmless if their operations and capacities can be explained by invoking less sophisticated internal sub-sub-systems, which can then be subsequently explained by appealing to even dumber internal sub-sub-sub-systems, and so on, until we just have brute physical mechanisms. Dennett's proposal is that this same decompositional strategy can work to "discharge" homunculi that serve as internal users of mental representations. Consequently, invoking the same type of arrangement that gives rise to non-mental representation – that is, representation plus an interpreter – is not such a problem after all.

To some degree, this decomposition strategy is the way the IO notion of representation, discussed in chapter 3, gains explanatory purchase. Inner sub-modules engaged in computational operations require symbolic representations of the inputs and outputs of the function they are computing. Adders can't be adders without using representations of numbers as input and producing representations of sums as output. Yet the invoking of an internal adder raises no regress concerns because an adder can be mechanically decomposed. So Dennett's story about discharging inner representation-users fits the way this more technical notion of representation answers the job description challenge. With this notion of representation, the invoking of a representation-user needn't lead to a vicious regress.

However, it is far from clear that this strategy can be extended to the receptor and S-representation notions. There are significant differences between the way we, on the one hand, use smoke to find out about fires, or a map to learn about some terrain, and computational sub-systems, on the other hand, use input–output symbols. Computational sub-systems use input symbols to designate the inputs to the function they are computing, and output symbols to designate the output value. For an adder, input symbols designate the numbers to be added. They are not used by the addition module to discover new facts, or to make inferences about further conditions or states of affairs. By contrast, when we use an external nomic dependent for representational purposes, or elements of a map or model, we do so to learn something new, often by making inferences from what the representation is telling us. We use the presence of smoke as a reason to infer the existence of a fire; we use the bend in the line on the map to conclude there is left bend in the path up ahead. For us, these uses involve the acquisition of new mental representations – the belief that something is on fire; the belief that the path turns left.

Hence, it would seem that for an inner state to be used in *exactly* the same way we use external signals, maps, models, and the like, the inner sub-system would need to do some fairly sophisticated mental operations, like using them to *learn* about something else. The homunculus decomposition strategy Dennett advocates only works if we have a good sense of how the relevant cognitive task can be carried out by a system that lends itself to functional analysis. But at the present, we don't have a clear understanding of how something as sophisticated as interpreting a symbol or acquiring new knowledge could be achieved by decomposable, mechanical sub-systems. In other words, the way *we* use external analogues to the receptor and S-representation notions is considerably more sophisticated than the way computational sub-systems use IO-representations. But the problem isn't just a matter of sophistication. It is also the case that the regress worry looms larger because, as we just saw, our everyday use of external representational devices involves the adoption of new beliefs about whatever it is the representation represents. If computational systems used inner representations in the same way, then this would also involve the acquisition of new internal representations, which would then require their own users, and we are off on a regress. Consequently, the homunculus-decomposition strategy Dennett proposes for handling the regress challenge is much less promising if we want an account of mental representation that is based on our ordinary use of receptor-style or S-representation-style external representations.

Thus, a third strategy for handling the regress worry has become quite popular. This strategy has been to offer an account of representation that drops the sophisticated homunculus, but nonetheless retains enough of the relevant relations, conditions and properties of representation so that, in the end, we still have something that is recognizably a representational system. In other words, the strategy is to show that we can have something that functions as a representation in a physical system, even if there is no sophisticated built-in learner or inference-maker that it serves as a representation *for*. Although many accounts adopting this strategy still appeal to the idea that there is a representation "consumer" (Millikan 1984), such a consumer is little more than a mechanical process or device that the representation effects. In fact, in some accounts the consumer is nothing more than the entire system itself. I'll refer to this as the "mindless strategy" for avoiding a regress. With this method, the way to understand mental representation involves coming up with an account of representation that requires nothing playing the role of an inner mind that uses the representation.

Now there may be accounts of representation in which it is far from obvious which of these different strategies is being pursued. There are, no doubt, some systems where the internal elements influenced by an alleged representational state are such that it is unclear whether they are supposed to be analogous to *us*, when we use these sorts of representations, or are instead primitive processors that facilitate an account of mindless representation.[1] Yet I believe that most writers attempting to explain cognitive representation by appealing to either the receptor or S-representation notions have done so by pursing a strategy that is clearly on the mindless end of the continuum. Indeed, what makes these two notions appealing as candidates for understanding mental representation is precisely the fact that they come to serve as representations *in part* because of properties they possess independently of any interpreting mind. The property of being nomically dependent upon some other condition and the property of sharing a structural isomorphism with other things are both properties that are sufficiently natural to suggest the possibility of constructing an account of representation that is properly mindless. That is one of the reasons why these two representational types have played such an important role in the way philosophers and cognitive scientists now think about cognitive representation. Yet, as I hope to make a little clearer, the mindless strategy does not work equally well for both of these representational notions.

In chapter 3, we saw that the S-representation notion is, indeed, an explanatorily valuable notion of cognitive representation – one that sits well with our basic understanding of what it is for something to function as a representation, even when the system lacks a sophisticated representation user or representation interpreter. In chapter 4, we saw that the same does not hold for the receptor notion. It fails as an explanatorily useful posit because it fails to assign a functional role to internal states that is recognizably representational in nature. If we re-frame this discussion in terms of the mindless strategy, we can say that the S-representation notion provides

[1] This point was made to me by Jaegwon Kim.

us with a form of representation whereby a physical structure can play a representation role even within a mindless system, whereas the receptor notion does not provide us with such a form of representation. Yet our earlier analysis may have left some readers unconvinced, particularly since the cases discussed were so dissimilar. A more direct comparison of the two notions, where they are employed by similar systems engaged in similar tasks, will allow us to see more vividly how the removal of a sophisticated cognitive agent from the picture undermines the representational status of the receptor state but not the S-representation state.

Suppose that we have three cars, A, B and C that manage to make their way along a segment of a track. The track has high walls, like a race track, and the segment they manage to navigate is shaped like a large "S." We can thus call it the "S-curve." Now suppose that the cars A and B have drivers, yet the cars themselves lack windows and thus the drivers cannot actually see where they are going. Car C, on the other hand, lacks any sort of driver. All three cars, however, manage to get through the S-curve successfully. Our goal is to explain how the cars manage to do this. After investigating the situation, we discover that the cars succeed in this task by using very different strategies. In car A, the driver uses a receptor-type system that works as follows. The car has rods that extend out at an angle from the corners of the car's front bumper. When one of the rods is pushed inward, as happens every time the car approaches one of the track walls, this closes a circuit and illuminates a light inside the driver compartment. There are two lights, one corresponding to each rod, and the lights are on the same side of the dash as the rod to which it is connected. For example, an illuminated right-hand light informs the driver that his car is approaching the right wall, causing the driver to steer the car in the opposite direction. Using this representational system, the driver manages to get through the S-curve quite well. By contrast, Car B employs an S-representational process. The driver has in her hands an accurate map of the S-curve. This is used, along with some sort of dead-reckoning strategy, to guide the driver through the S-curve without ever touching any of the walls. Finally, Car C lacks any sort of driver. It gets through the curve only by careening and bouncing off of the walls in an unguided manner. Every time Car C hits one of the wall, the wheels bounce in the opposite direction and this turns the car gradually away from the wall, toward the opposite wall. Car C operates in a manner no different than a marble in a pinball machine; its course is determined only by its brute interactions with the physical structure of the path it is on (see figure 6a).

The first two cars clearly use different representational systems that are exploited by interpreting minds, provided by the drivers. The different

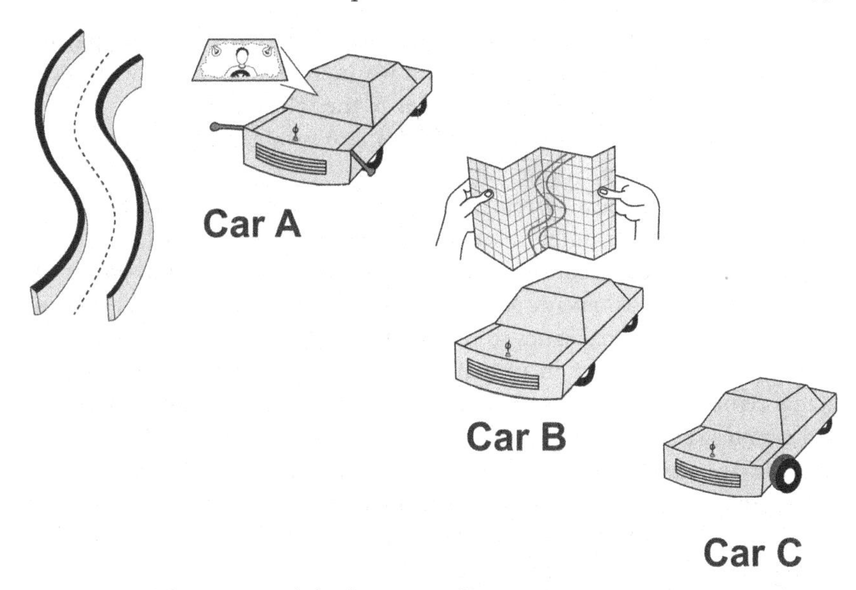

Figure 6a The S-curve and the three types of cars, with the drivers in Cars A and B using different representational strategies.

strategies used by these cars are familiar to submariners, who sometimes maneuver through underwater canyons by using sonar (analogous to the receptor strategy used in car A), and sometimes by using detailed maps and dead-reckoning (analogous to the S-representation strategy used in car B). The third car, on the other hand, clearly uses no representational structures. It gets through the S-curve by brute force. What we want to know is this: What happens when we remove the drivers from cars A and B and make these cars into mindless systems? We can imagine reconfiguring their internal workings in a way that allows the cars to negotiate the S-curve using much of the same basic apparatus that played a representational role when a driver was present. When the cars are modified in this way, do their inner states and structures retain their status as functioning representations? Or do they instead make the car more like car C – careening through the curve without using internal representational states? My claim is that car A – the one using the receptor-style state – stops being a representational system, whereas car B – using S-representations – retains its status as a representational system, despite being mindless. Seeing why and how this is so will help us better understand why the one notion of representation belongs in cognitive theories while the other notion does not.

Before we see what happens when we remove the drivers, it will pay to remind ourselves of the sort of question we are trying to answer with such an investigation. The question is neither of the following:

(a) Are there mindless systems for which it is possible to characterize their inner processes in representational terms?

(b) Are there mindless systems for which it is absolutely necessary to characterize their inner processes in representational terms?

Neither of these questions is worth pursuing because they both admit of trivial answers. It is always *possible* to adopt the intentional stance and ascribe representational states to just about anything, including rocks and vegetables. And, at least in principle, it is never *necessary* to characterize actual representational systems in representational terms because all physical systems can be described in purely causal–physical terms. Instead, the question we are interested in answering is something more like this:

c) Are there mindless systems in which an internal element is performing a role that is most naturally (or intuitively, or justifiably, or beneficially) viewed as representational in nature?

As I pointed out in chapter 1, a question like this is not as crisp as we would like, in part because it invites us to make a judgment call. Nonetheless, I believe the question can be answered in a way that is illuminating and important to our understanding of representation in cognitive theories. The correct answer is "yes," but only if we use the right sort of representational notion. To see this, consider what happens when we remove the drivers from the cars.

In the case of car A, we can easily imagine a modification whereby we replace the light with some sort of relay that activates a servomechanism whenever the rod is pushed inward. When engaged, the servomechanism steers the wheels in opposite direction of the plunged rod, as roughly illustrated in figure 6b. The result is a vehicle that behaves in a manner not unlike the way it used to behave when there was a driver behind the wheel. As it approaches one of the walls, the relevant rod is plunged and the servo-mechanism is engaged, turning the vehicle away from the wall and allowing it to move through the curve. The important question, however, is whether the new steering process is one that is still representational in nature. I believe that it isn't. While there is indeed a causal process that uses relay structures that are supposed to go into specific states whenever a given condition obtains (one side coming close to a wall), there is really no sense in which any of these elements functions in a manner that is recognizably representational. When explaining how the mindless car A makes it way though the curve, the account that seems most natural (and fulfills our

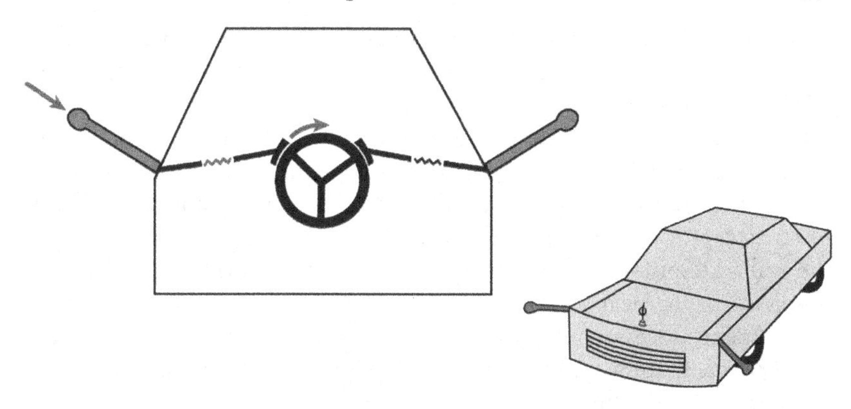

Figure 6b The modified, mindless version of Car A. A rod pushed inwards causes the steering wheel to turn in the opposite direction.

explanatory goals) is one that treats the causal relay between the plunged rod and the turned wheels as just that – a causal relay that brings about an altered wheel alignment whenever the vehicle gets close to a wall. In fact, there is no deep difference between this causal process and the one that changes the position of the wheels in car C. In both cases, we can explain this process as one of brute causal interaction between the wall and the wheels. True, the causal chain in the modified car A is here a little longer. There are certainly more mediating links between the car's proximity to a wall and the turning of the wheels away from the wall. But – and this is the key point – there is no natural or intuitive sense in which one of the linking elements is playing the role of representing. There is no intuitive sense in which the plunged rod, the closed relay circuit, the engaged servomechanism, or any of the other components in car A are performing a function that is representational in nature, any more than the rebounding wheels in car C. Car A is now bouncing off the walls too – it is just doing so in a more complicated fashion.[2]

[2] In fact, we can gradually morph car C into car A by increasing the complexity of the causal chain while never crossing a metaphysical point where the car changes into a representational system. Suppose the wheels of car C have protruding rods on their side, so it is actually these that hit the walls. The rods bump into the wall and thus push the wheels away, but they clearly do not represent. Now suppose the rods are actually connected directly to the top of the steering wheel. When plunged, they now push the steering wheel away from the wall, but without serving as representations. Now imagine we replace this arrangement with one where the plunged rod engages the servomechanism that actually turns the steering wheel. This is car A, but we have added nothing to the causal process that is functioning as a representation.

Suppose the causal chain is mediated by something that has been recruited because of its pre-existing tendency to reliably respond to the wall in some manner. For example, suppose the wall is illuminated in some way so that we can replace the rod assembly with a photo-receptor. The photo-receptor reliably responds to photons coming off the wall and in turn closes the relay circuit that drives the servomechanism. Does the photo-receptor now serve as something that represents the proximity of the wall? Despite the common tendency to assume that it would (to be discussed in the next section), it is again difficult to see why this sort of recruitment process would give rise to a representational system. One sort of causal mediator would simply be replaced with another. But this has no real bearing on the functional role of the mediator. It responds to photons in roughly the same general manner in which the rod responded to a nearby wall. Our understanding of the success of the vehicle is not enhanced by treating it or any other aspect of this process as standing for something else. Non-representational causal processes are still non-representational causal processes, even when one of the intermediary elements is put in place because it has the property of reliably responding to certain features of the environment.

Now consider car B. Suppose we remove its driver and automate its internal workings in a manner that also preserves at least some of the mechanics of the earlier representational system. One way we might do this, suggested by Cummins (1996), would be to convert the S-curve map into an S-shaped groove into which a rudder would fit. The rudder could then move along the groove as the vehicle moves forward, and the direction of the steering wheel and, thus, the vehicle's front wheels could be made to correspond to the direction of the rudder, as illustrated in figure 6c. As the rudder moves along the groove, its change in orientation would bring about a change in the orientation of the front wheels. Because the shape of the groove is isomorphic with the curve itself, the wheels change along with the S-curve and the vehicle moves through it without ever bumping into a wall.

What is the most natural and explanatorily beneficial way to describe this system? If we are just describing the internal mechanics of the system, as I just did, then we never really need to treat any of the elements as representations. We can avoid representational and map language when describing the system if we try hard enough, just as we could when there was a driver in the vehicle. The physical stance is always possible. Yet if we want to explain how the vehicle actually manages to successfully navigate the S-curve, and we ask about the functional role the groove plays in this process,

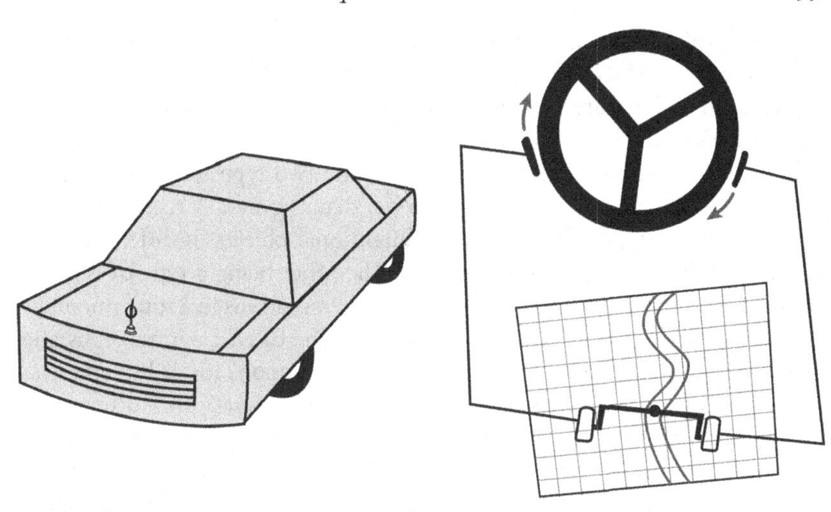

Figure 6c The modified, mindless version of Car B. As the rudder moves along the groove, the direction of the front wheels corresponds to the orientation of the rudder.

then the most natural and obvious thing to say is that the groove is serving as a map of the course of the track. After all, the car is exploiting the isomorphism between the groove and the track in much the same way that the driver did, even though the process is now fully automated and mindless. If we want to know how the vehicle manages to turn its wheels to the left at the very location it needs to, part of the answer would appeal to the leftward curve in the area of the groove that *stands in for* the corresponding area of the track. And to say that an area of the groove functions as a "stand in" for a segment of the track is just to say that an area of the groove is playing a representational role. Unlike the situation with the modified version of Car A, it takes considerable contrivance and effort to *not* view the modified version of car B as using a representational system. That's because the mindless version of Car B is significantly dissimilar to Car C and mindless Car A. Whereas the latter two cars proceed by using structures that facilitate a wall rebound process, mindless Car B has an internal element that is still used for what is most naturally viewed as a *guidance* process. It is still relying upon a dead-reckoning strategy, one that utilizes a component with the job of making the wheels turn in a manner that corresponds to the shape of the curve. That component, the groove, does that by modeling the curve. And the modeling of the curve is done by mirroring its form – by playing the role of a surrogate curve, with locations that stand in for locations along the real-world curve.

What explains this difference between the two sorts of cases? Why are we able to make an S-representation system mindless, and have it retains its representational nature, but we can't do this for receptor representation systems? My hunch is that in ordinary mindful scenarios (where we are using external representational structures), the two types of representation differ in a way that is more basic and deeper than is generally recognized. It is not only that they are linked to their intentional objects in different ways, one being a nomic dependency relation, the other being a type of isomorphism relation. It is also that they serve as representations in a fundamentally different manner. As we saw in chapter 4, in the case of natural signs and receptor-type states, these things serve as representations for us by serving as *informers*. Minds exploit the fact that certain states co-vary with other states of the world by making inferences and thereby acquiring new beliefs. We infer that because a given condition nomically depends upon (or reliably co-varies with) some other condition, the former reveals that the latter obtains. The role of the former state is to inform the cognitive agent about this other condition. But, of course, for something to work as an informer in this sense, there needs to be an *informee*. That is, there needs to be a cognitive agent who can recognize and exploit the natural dependencies to make the relevant inferences and thereby learn something new. Since becoming informed in this sense is quite a sophisticated cognitive operation, the role cannot be maintained without a fairly sophisticated representation consumer; that is, a fairly sophisticated cognitive agent. Structures stop functioning as representations once such a cognitive agent is removed from the picture, even though the same nomic dependency may be exploited in other ways.

By contrast, in the case of S-representation, structures serve as representations by serving as elements of a map, model, or simulation. The elements of the model function as surrogates – as stand-ins or substitutes for aspects of the model's target. In the mindful case, a cognitive agent exploits the similarity between the model and the target in order to make various inferences and learn things, just as with the receptor notion. But in this case, internal structures can continue to function as a model or map, and thus its elements can still serve as surrogates (as stand-ins), even if we drop the involvement of an inferring and learning mind. This is because surrogative problem solving can be automated. A mindless system can still take advantage of the structural isomorphism between internal structures and the world, and in so doing, employ elements of those internal structures as representations-qua-stand-ins.

Thus, in the receptor case, we have a nomic (usually causal) dependency relation that can be exploited by minds to infer one set of conditions on the basis of another set of conditions, and it is this sort of inference that amounts to a sort of interpretation and bestows upon the latter the role of representation. The notion of representation at work is *not* one of something standing *in* for some thing or condition, but *implying* or *entailing* some condition. The inference is what converts this mere entailment relation (based upon the dependency) into a representational relation. In order for the property of nomic dependency to give rise to the property of standing for something else, we need the involvement of an inference-maker. Without a mind making the relevant inference, the dependency relation can still be put to use (perhaps to make something into a reliable causal mediator), but not for representational purposes. On the other hand, in the case of S-representation, we have an isomorphism relation that can also be exploited by minds to make inferences from one set of conditions to another. Yet in this case, a state serves as a representation not simply by virtue of such an inference, but also by virtue of functioning as a state that stands *in* for something else, in the sense of serving as a surrogate. This surrogative, "standing-in-for" property is not dependent upon any inference or learning process because, as we've seen, inner structures can still serve as elements of simulations, models or maps even when no independent mind treats them as such. Surrogative reasoning can be, in this sense, mindless, and thus the surrogates themselves can serve as representations in mindless systems.

These deep differences in the way structures serve as representations – representing X by entailing X versus representing X by standing in for X – do not produce the same results when we incorporate these ideas into cognitive theories. Therefore, it is important not to conflate the two notions. In the next section, we will look at accounts of representation that fail to properly distinguish the two notions and thereby fail to present an accurate picture of various complex systems. But first, it is worth considering another notion of representation in terms of our hypothetical cars. In the last chapter, we saw that theories that posit tacit representational states are just as misguided as those that posit receptor representations because there is nothing even remotely representational about the roles played by the dispositional states identified by these theories. Although it is harder to find non-mental analogues for the tacit notion, we can still frame our critical comments of this notion in terms of the S-curve example. Strictly speaking, Car C – the one that lacks any overt representational device – has a functional architecture that enables it to negotiate the S-curve. It has the dispositional property of being able to get through the curve when the right conditions are met. If we adopted the same criteria for assigning tacit representations that it seems many writers adopt, we could say that Car C has tacit "know-how" with respect to navigating the curve. Although Car C needs to interact with the environment (the walls of the curve) to succeed, this shouldn't matter given the way writers have invoked the notion of tacit representation in the past. After all, Dennett's chess-playing device requires input to succeed as well. So if we take seriously the tacit notion, Car C is a representational system after all. But this is clearly absurd. In all relevant respects, Car C is no different than a boulder rolling down a hill that bounces off of various obstacles. While the dispositional nature of Car C's front wheels explains how the car rebounds off of the walls in a manner that gets it through the curve, the wheels (or their dispositional nature) serve no representational function.

Now suppose we alter the functional architecture of Car C in a manner that makes the car's behavior more dependent upon its inner workings and less dependent upon interactions with the environment. We could presumably do this by installing some sort of internal mechanism that shifts the direction of the wheels from side to side as it moves forward, perhaps with an elaborate set of gears or maybe just a timing device. There would be no individual item that is functioning in the manner of the groove that is employed in the modified version of Car B, but there would still be something about the functional design of the system that causes the wheels to rotate back in forth in a way that allows it to move through the S-curve successfully. To make the situation even more analogous to cognitive systems, we can pretend this functional architecture somehow resulted from an earlier trial and error period that involved attempts to navigate the curve – a process some might be tempted to call learning. What would this do to our analysis?

I'm quite willing to allow that as the functional architecture is transformed in various ways, we may get to a point where there is something at work that looks more and more like a representational system. Since I think the representational status of any system depends to some degree on its correspondence with our ordinary understanding of representation, a judgement call is needed. Moreover, I agree with others who have suggested that representation might come in degrees – that it is not necessarily an all-or-nothing affair (see, for example, Clark and Toribio 1994). But it is important to realize that even if we grant that the modification of Car C is now starting to move it into the realm of an intuitive representational system, this is happening not because the tacit notion is being made legitimate, but rather because the transformation is moving us closer to something like a functioning form of S-representation. We are inclined to treat the car with the proposed gear device as a representational system because it is possible to view the gear assembly as serving as a model of the curve, whereby, say, specific positions of the gear rotation actually stand in for points along the S-curve. This, of course, would depend on the details. The critical point is that this would not be a case where the tacit notion of representation would be vindicated. Instead, this would be a case where the car's internal design is modified so that it starts to look like the car is using something like an internal model, and thus S-representations, to get through the curve.

The point of this discussion has been to offer a more direct comparison of the two most significant notions of representation in cognitive science, looking at how they might be employed to explain similar physical systems. I've framed this comparison in terms of a certain strategy for introducing and explicating representational posits – one that involves taking familiar types of representations that we use "out" in the world and attempting to fashion an account of cognitive representation from these. I've done this not just because I think that this is how most philosophers and researchers actually do try to explain representation in cognitive systems, but also because it sheds considerable light on how and why cognitive theories appealing to one form of representation make sense, whereas accounts appealing to the other form involve a conceptual confusion. But this confusion is almost never recognized, and recent attempts to defend a representation-based cognitive science from an anti-representational movement tend to reiterate this mistake by offering variations on the receptor theme. It will help our understanding of these matters if we take a look at this debate.

6.2 DYNAMIC SYSTEMS THEORY AND THE DEFENSE OF REPRESENTATIONALISM

Throughout my analysis I've suggested that cognitive theories today are ostensibly committed to inner representations, even though a closer analysis very often reveals that states characterized as playing a representational role aren't doing any such thing. But this characterization of contemporary research is slightly misleading. While the overwhelming majority of theories appeal to a notions of inner representations, a small group of researchers, especially in the field of robotics, have attempted to develop a radically new approach to understanding cognition that self-consciously avoids the positing of inner representations. While these iconoclastic theories vary in their aims and explanatory goals, they are often grouped under a general heading called "Dynamic Systems Theory" or DST for short. DST is often treated as offering a radical new framework for understanding cognitive processes, not just because it typically eschews invoking inner representations, but also because it seeks to replace much of the theoretical machinery of both traditional computationalism and connectionism with more abstract mathematical analyses of a cognitive systems. In response to this research, several philosophers have developed sophisticated defenses of representationalism that are designed to show that inner representations are far more prevalent and far more indispensable than DST proponents recognize. This philosophical defense can also be seen as providing a response to my own doubts about representation – especially those presented in chapter 4 and in the last section of this chapter. Thus, a closer look at these pro-representation accounts is needed to see if my earlier analysis has been unfair. Before we do this, however, it will help to briefly consider some of the theoretical work that has prompted this debate.

6.2.1 Dynamic anti-representationalism

Although it is not presented as an example of DST as such, the robotic research of Rodney Brooks displayed in his well-known "Intelligence Without Representation" (Brooks 1991) is often treated as a precursor to cognitive applications of DST since his account embodies many of the same principles. Brooks's perspective on cognition is driven, in part, by frustration with what he sees as the contrived manner in which many traditional AI researchers define problem domains by specifying neat but artificially simple sub-tasks. Instead, Brooks has developed robots that move about and engage in various search-oriented tasks in the real world. His robotic devices do not use internal modules designed to perform various sub-routines, nor a central control mechanism that utilizes an array of internal representations. The architecture of his robots is organized so that complete skills, such as the capacity to move without running into objects, are "layered" on top of one another. In other words, a relatively non-modular system that enables the robot to move and avoid objects serves as the foundation for another, more complex system designed to seek out specific targets. While various override mechanisms intervene when the goals of the two systems conflict with one another, there is no central processing system or set of centrally accessible representations that guide the robot's overall behavior. According to Brooks, this approach leads to two radical claims about intelligent systems:

(C) When we examine very simple levels of intelligence we find that explicit representation and models of the world simply get in the way. It turns out to be better to let the world itself serve as its own model.
(H) Representation is the wrong unit of abstraction in building the bulkiest parts of intelligent systems. (1991, p. 139)

Presumably, by the phrase "the world itself serves as its own model," Brooks means not that the real world somehow functions as a model, but rather that his robots directly interact with the world without using an intermediary representational model. Thus, his robots are designed to perform various tasks using a complex internal architecture, and this architecture helps the system move about the environment. However, according to Brooks, there is nothing within this architecture that actually guides the system by standing for different aspects of the environment.

Another investigator who develops relatively simple robotic systems and also questions the need to posit internal representations is Randall Beer (Beer and Gallagher 1992; Beer 1995). The sort of phenomenon that Beer focuses upon – namely, relatively simple systems dealing with real-world complexities in real time – is commonly characterized as investigating "embodied" or "embedded" cognition. While much of Beer's work involves simulations rather than real robots, he nevertheless shares with Brooks a common emphasis upon the way so-called "autonomous agents" interact with a real-world environment. Yet Beer is much more explicitly committed to DST as his central explanatory framework. For Beer, the agent–environment interplay is treated as involving two coupled dynamic systems that move through various phases in an abstract state-space, all in accordance with various dynamical laws. The state–space trajectories and positions of this complex system are described using the vocabulary of DST, designating the unique theoretical posits. For example, if the trajectory of a dynamic system tends to repeatedly fall into and occupy an area of phase-space, this area is called an "attractor." Other dynamic notions include "repellors," "equilibrium points," "separatrices," and "phase portraits." Beer argues this framework provides an explanatorily and predictively successful theory about the behavior of cognitive systems in complex and changing environments. While the details of Beer's account needn't concern us here, it is worth emphasizing that his account shares my own skepticism about equating representation with just causally significant internal states or components of a cognitive system. Echoing themes presented in earlier chapters, Beer worries that if we adopt a weak notion of representation proposed by many, then representationalism would "make no interesting theoretical claim about the nature of the processes underlying behavior ... representation must require additional conditions above and beyond the mere possession of internal state (or correlation, etc.)" (1995, p. 211).

The DST approach has also been used to provide a new outlook on various dimensions of cognitive development. Perhaps the most detailed account is offered by Esther Thelan and Linda Smith (1994). These authors argue that conventional categories for understanding development – nativist, empiricist, constructionist – all are fundamentally wrong-headed. One of the central problems with these approaches, they claim, is that they assume development is guided by some sort of inner blue-print (either in-born or acquired through experience) that serves to direct a child's cognitive growth. Instead, cognition should be understood as a dynamic process driven by, as they put it, "time-locked patterns of activity across heterogeneous components" (1994, p. 338). To establish their position, Thelan and Smith take a close look at the different phases children go through while learning to walk. They insist that this form of learning is best understood not as the unfolding of an internally represented succession of motor skills, but rather as a trajectory through a dynamic attractor landscape, the nature of which is determined by various complicated organism–environment interactions. More central for our discussion is their insistence that the proper framework for understanding cognitive development will abandon the old notions of psychology, especially those associated with mental representations. As they put it, their theory "suggests that explanations in terms of structures in the head – 'beliefs,' 'rules,' 'concepts,' and 'schemata' – are not acceptable; acceptable explanations will ground behavior in real activity. Our theory has new concepts at the center – nonlinearity, re-entrance, coupling, heterochronity, attractors, momentum, state spaces, intrinsic dynamics, forces" (1994, pp. 338–339). While the jargon in their exposition may leave readers unclear on exactly what DST says cognition is, it is fairly clear from their account what it isn't, namely, processes driven by internal representations.

Of all the challenges to representationalism associated with DST, however, none is as detailed or has proven as provocative as the account presented by van Gelder (1995). Van Gelder's defense of DST is framed around his discussion of a simple mechanical device designed by James Watt in 1788 to regulate the functioning of steam engines. To maintain a constant engine speed, the Watt governor uses a simple but ingenious feedback mechanism illustrated in figure 6d. Van Gelder provides a good description of its operation:

> It consisted of a vertical spindle geared into the main flywheel so that it rotated at a speed directly dependent upon that of the flywheel itself. Attached to the spindle by hinges were two arms, and on the end of each arm was a metal ball. As the

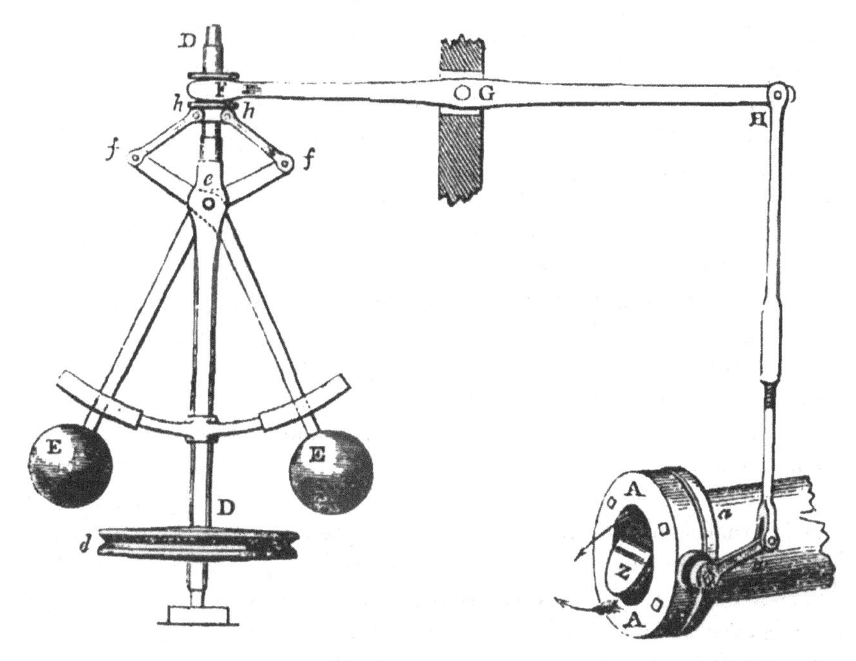

Figure 6d The Watt Governor. From *A Treatise on the Steam Engine*, John Farey, Longman, Rees, Orme, Brown, and Green (1827).

> spindle turned, centrifugal force drove the balls outward and hence upward. By a clever arrangement, this arm motion was linked directly to the throttle valve. The result was that as the speed of the main wheel increased, the arms raised, closing the valve and restricting the flow of steam; as the speed decreased, the arms fell, opening the valve and allowing more steam to flow. The engine adopted a constant speed, maintained with extraordinary swiftness and smoothness in the presence of large fluctuations in pressure and load. (1995, p. 349)

Van Gelder contrasts this device with a hypothetical computational governor – one that utilizes a number of sub-systems doing things like measuring the speed of the flywheel, calculating the difference between actual and desired speed, measuring steam pressure, and so on. These different functional components could then be coordinated by a central executive system that manipulates representations of different aspects of the problem domain. Thus, the hypothetical governor would operate by using the same sort of elements and principles found in classical computational systems. The point of the comparison is to illustrate how engineering problems that might be assumed to require a computational solution can be handled instead by non-computational, purely mechanical devices. The actual Watt governor uses no elaborate sub-systems and, according to van Gelder, no internal representations. How then should its operations and functional architecture be characterized? Van Gelder offers a detailed DST analysis, one that treats the equilibrium state reached between the governor's weighted arms and the proper engine speed as an attractor. According to him, DST provides the most useful explanatory framework for understanding a subtle dynamic system like the Watt governor.

The primary goal of van Gelder's exposition is to suggest that cognitive systems, like our own brains, should also be treated not as computational or connectionist systems, but as dynamic systems. Such a radical shift in cognitive theorizing would entail a host of changes; yet van Gelder indicates that the most significant modification would be an abandonment of representational posits. While he admits that DST is not inherently anti-representational in nature, he offers a number of reasons against treating the internal elements and states of dynamic systems, like the weighted arms of the Watt governor, as representations. For example, he points out that describing the arms in representational terms provides no explanatory payoff, and no one who actually describes the functioning of the governor does so in representational terms. Despite the fact that the position of the arms is (more or less) correlated with the speed of the engine, van Gelder rejects the notion that this is sufficient for treating the arms as representations. Moreover, because "the arm angle and engine speed are at all times both determined by, and determining, each other's behavior," their relation is, according to van Gelder, "much more subtle and complex" than can be captured by the notion of representation (1995, p. 353).

A good deal of what van Gelder says about representation overlaps with my own treatment. He is correct to challenge the explanatory utility of treating internal structures as representations just because they are causally implicated in the processing, or because they co-vary with specific types of input to the system. In fact, the Watt governor serves to illustrate many of the same points made in the last section by the mindless version of Car A in the S-curve. In both cases, a proper analysis of the physical system's design reveals components that share *some* of the features accompanying a type of external representation (like the need to go into certain states only when certain conditions obtain), but not enough to justify treating these components as actually playing a representational role. In both cases, we see how adopting very weak criteria for representation would give rise to an embarrassing over-proliferation of representation structures and systems. What's missing from van Gelder's account is a functional analysis that assigns to the weighted arms a proper job description and thereby reveals *why* they shouldn't be treated as representations. They shouldn't be treated as representations (and doing so would have no explanatory utility) because they actually serve as a type of causal mediator; their function is to convert changes in engine speed into changes in valve position. They don't serve to *measure* or *record* or even *monitor* engine speed, but they do serve to mechanically *transform* specific speeds into specific valve-flap settings. This is an important engineering role from the standpoint of solving the governing problem; but it is not a role that makes something a representation.

6.2.2 The representationalists strike back

As might be expected, the anti-representational nature of the DST framework has generated a strong back-lash, especially from philosophers committed to preserving the representational paradigm (Bechtel 1998, 2001; Clark and Torbio 1994; Clark 1997; Chemero 2000). Many of these counter-arguments strike me as plausible rebuttals to at least some of the claims made by DST advocates. For example, a number of writers point out the degree to which dynamic system theorists have overstated their conclusions. One common sort of DST argument relies upon a non-sequitur: Because robots and other systems can perform fairly simple tasks without internal representations, representations are therefore unnecessary for any cognitive task, no matter how sophisticated. This conclusion is clearly unwarranted since the simple robotic systems that Brooks and Beer develop don't tell us much about more sophisticated tasks, such as those that require careful planning or hypothetical reasoning (Clark and Torbio 1994; Clark 1997). Moreover, a number of representation defenders also note that it is not clear that the sort of framework DST provides is an alternative *explanatory* framework, as opposed to a more abstract and mathematical *description* of the behavior of complex systems (Chemero 2000). Since virtually every complex system lends itself to dynamical characterization, in one trivial sense the brain is a dynamical system *even if* it is also a classical computational or connectionist system as well. DST may provide a new and perhaps useful vehicle for re-framing the way we describe cognitive phenomena, yet also be compatible with computational explanations of the mechanisms of the mind. It is perhaps worth noting that in van Gelder's useful explanation of how the Watt governor works, he talks about good-old-fashioned inner causal mechanisms like flywheels, spindle arms, and shut-off valves, *not* dynamic posits such as attractors, repellors, phase points and the like. In fact, had he offered only the DST account of the governor's operations, I suspect most readers would be left scratching their heads, still wondering how the system *actually works*.

While these and several other points made by the defenders of representation strike me as perfectly reasonable, another line of defense adopts a strategy that I believe is seriously flawed. This line of defense often focuses on the claims made about dynamic systems like the Watt governor, and typically involves three major steps. The first step is to offer a definition or analysis of representation that broadens the range of structures that qualify as "modest" representational states, beyond, especially, the robust symbolic states found in the CCTC. The second step is to show how this definition allows us to treat internal states of simple systems as representational after all. The third step is to defend this analysis by noting how it accords with the theories and representational posits that often appear in the neurosciences and connectionist research. Because I have been critical of both the representational posits in these sorts of theories, and of the weaker, more inclusive conceptions of representation, this line of response to DST can be seen as a direct challenge to my account and thereby warrants a closer look.

Perhaps the most ambitious defense of representationalism is offered in a couple of papers by William Bechtel (1998, 2001). Bechtel begins with an analysis of representation inspired by Newell's analysis of designation: "An entity X designates an entity Y relative to a process P, if, when P takes X as input, its behavior depends on Y" (Newell 1980, p. 156). As we noted in chapter 1, it is not immediately obvious just how we are to unpack this definition, but Bechtel suggests that we should understand it as implying that an internal structure serves as a representation if it "stands in" for other things, like aspects of the environment. Bechtel embraces van Gelder's analysis of representation (which is itself based upon suggestions from Haugeland 1991) that cognitive representations stand in for some further state of affairs, "'thereby enabling the system to behave appropriately with respect to that state of affairs'" (Bechtel 1998, p. 297).

So it initially appears that Bechtel (like Haugeland and van Gelder), adopts the specific conception of representation associated with the S-representation notion: internal structures serve as representations by serving as stand-ins or surrogates in some broader model, simulation or map. However, it soon becomes clear that Bechtel's conception of the "stand-in" relation is very different than the one I've offered here. To explicate what it means for a state to stand in for something else, Bechtel appeals to Dretske and Millikan's account of receptor-style representation in which a given state represents because it has the function of co-varying with some other state of affairs. In other words, Bechtel runs together the two notions of representation we have distinguished – representation-*qua*-surrogate and representation-*qua*-nomic-dependent – and treats these as aspects of a single, unified conception. For him, any state whose functional role is to reliably respond to some other condition is thereby viewed as standing *in* for that condition, and thus serves as a representation.

With this analysis of representation in hand, Bechtel returns to van Gelder's Watt governor example and insists that the governor actually does employ internal representations. For Bechtel, the weighted arms attached to the spindle,

> stand in for the speed of the steam engine in a mechanism that controls the opening and closing of the steam valve and it is *because* they so stand in that the governor is able to regulate the flow of steam appropriately. If someone did not immediately perceive how the governor performed its function, what one would do is explain how the angle of the arms responds to the speed with which the flywheel is turning (i.e., it carries information about the speed), thereby making this information available in a format that can be used by the linkage mechanism to control the flow of steam and the speed of the flywheel at a future time." (2001, p. 335)

Thus, Bechtel insists that van Gelder is just wrong to suggest that a proper understanding of the governor will not appeal to internal representations. Indeed, Bechtel goes on to note that it is precisely this notion of representation that is so prevalent in the cognitive neurosciences, as illustrated by Snyder's account of visual processing and attention (Snyder *et al.*, 1997). In these theories, neural states that fire in response to stimuli, and that fire to cause movement, are generally treated as playing a representational role.

Bechtel's argument can be seen as using the following conditional: If his (Bechtel's) account of representation is correct, then the internal arms of the Watts governor are serving as representations. I accept the conditional, but, as the saying goes, one person's *ponens* is another person's *tollens.* Actually, "tollensing" in this case is more than a matter of personal preference, and given our earlier analyses, it should be immediately clear where Bechtel's analysis goes off the tracks. The mistake is in conflating an explanatorily useful notion of internal representation (representation-*qua*-surrogate) with something that really isn't functioning as a representation at all (a reliable causal mediator). According to Bechtel's analysis, when there is an internal state B that is caused by some other state A, and A's causing B brings about the desired operation of the system, then B is *standing in for* A. So on Bechtel's account, a stand-in representation is nothing more than an internal element that forges a causal link between certain conditions and suitable output. "Standing in" for something just means "affected by" that thing in a useful and reliable way. This is a radically different sense of the "standing in for" property than was presented in my analysis of S-representation. As we saw in chapter 3, to serve as a stand-in is to serve as a functioning surrogate or substitute that is a component of some sort of model or simulation of a target. Bechtel's account makes no mention of this sort of surrogative functional role. Instead, it treats the weighted arms as stand-ins for engine speed simply because they serve to link the engine speed to a process of engine regulation.

Yet it is hard to see why the sort of basic causal relation that Bechtel describes should be characterized as one whereby one thing stands in for something else. Do spark plug firings stand in for accelerator pedal positions, or does the dripping coffee stand in for my turning on the coffee-maker? If they do, it is only in a very uninteresting sense that reflects nothing more than basic causal relations. To serve as a "stand-in" in *this* sense is to serve not as a representation, but as a reliable causal bridge. To help see the sort of problems Bechtel's analysis of representation would generate, consider that on his account many more components of Watt's governor would play the role of representation than just the weighted arms. In fact, on his conception, virtually every causally relevant element of the system between the steam engine and the valve serves as a representation. The angle of the arms "stand in for" the speed of the engine, but then so does the speed of the flywheel, the speed of the spindle, the angles of joints *f* and H in the diagram shown in figure 6d, the height of the coupler F, and so on. All of these elements go into states that depend upon the engine speed, and all of these are specifically designed to have a causal influence on the valve flap. In fact, many of these elements would actually qualify as playing a triple role, serving as a state that is represented, a representation of some other state, and a representation user, *all at once*. For instance, the angle of the arms, is, in Bechtel's sense, represented by the height of coupler F; it also represents the speed of the flywheel; and it also qualifies as a user of a representation of engine speed that is instantiated by the flywheel! So on Bechtel's account of representation, virtually every internal element of the Watt governor qualifies as playing at least some of the roles associated with representation. Moreover, this excessive proliferation of representation would not only infect our understanding of the Watt governor, but also virtually every complex system that is produced by some sort of design process.

Like many writers who endorse the receptor notion, Bechtel's functional analysis of the weighted arms is clouded by his adoption of informational language: "the angle of the arms responds to the speed with which the flywheel is turning (i.e., it carries information about the speed), thereby making this information available in a format that can be used by the linkage mechanism . . ." (2001, p. 335). As we noted in chapter 4, in a weak and uninteresting sense, every reliable effect can be said to carry information about its cause. The interesting question is whether something is *functioning as* an information carrier, and doing so in a stronger sense than functioning as a mere reliable respondent. The arms of the governor could, of course, function as information carriers in the stronger sense, say if *we* used their position to ascertain the speed of the flywheel. But, to repeat the points made in prior chapters, just because something *could* be used as information carrier in this stronger sense, it doesn't follow that it actually is. In the case of the Watt governor, the weighted arms serve not as information carriers in this stronger sense, but as a way of forging a causal link between the speed of the engine and the opening and closing of the steam valve. If someone didn't understand how the governor worked, we would not, as Bechtel suggests, describe the arms as information carriers. Instead, we would do exactly what van Gelder does, namely, explain how the speed of the engine causes the spinning spindle, which in turn causes the arms to lift from centrifugal force. Then you would show how this brings about the opening and closing of the steam valve through the purely mechanical operation of the attaching mechanisms. It is not only unnecessary to invoke talk of information or representational stand-ins, it is also wrong to do so. There really is *no* interesting sense in which the weighted arms play the functional role of carrying information about something else.

Bechtel is aware that his account at least implies an over-proliferation of representations and representational systems, and thereby suggests the role of mere causal mediation should be supplemented by the inclusion of a representation *consumer*. Initially, this looks like a promising way to strengthen the account to yield something closer to our ordinary understanding of representation. Yet on Bechtel's theory, a representation consumer is nothing more than an element of the system whose own states are causally influenced by the alleged representation. Bechtel explains, "Representations are found only in those systems in which there is another process (a consumer) designed to use the representation in generating its behavior (in the simplest case, to coordinate its behavior with respect to what is represented) and where the fact that the representation carried that information was the reason the down-line process was designed to rely on the representation" (2001, p. 337). This certainly sounds good, until we unpack terms like "consumer" and "carried information." When we get clearer on what all of this really means, as shown by his analysis of the Watt governor, Bechtel's statement admits of the following translation: *Representations are found only in those systems in which there is another process (a consumer) designed to be causally influenced by an internal state whose own behavior is itself caused by a prior state and where the fact that the internal state reliably responds to that prior state was the reason the down-line process was designed to be influenced by the internal state.* In other words, representation is said to occur whenever there is some sort of internal causal mediator. The invoking of a consumer (in this sense) does nothing to enhance the role of the state beyond that of causal relay.

Consequently, Bechtel's analysis is a classic example of what we described in chapter 1 as succumbing to the Charybdis of "over-naturalizing" representation; that is, offering criteria for representation that are so mechanically mundane that they make representation into little more than a causally relevant internal state. His account thereby fails to capture our intuitive understanding of what a representation is and does. Ultimately, we want to know how a physical system S can have an internal state X that serves to represent Y, perhaps for consumer C. But on his analysis, this involves nothing more than S being designed so that Y causally influences some process C via an intermediary state X. In an effort to preserve an explanatory role for representation, Bechtel has reduced representation to nothing beyond simple causal mediation. While we want a naturalistic, reductive account of representation (one that explicates how representation can occur in causal–physical terms), we don't want a reductive analysis that simply abandons our intuitive understanding of representation altogether. When we ask if a given physical system is using inner representations, we are not simply asking if the system is using internal states that function as causal go-betweens.

There are, however, two important elements of Bechtel's account that I believe are correct. First, Bechtel is correct to reject one of van Gelder's main arguments against a representational treatment of the governor's components. Van Gelder insists that the weighted arms can't be representations because their relation to the engine speed involves a feedback relation that is too "subtle and complex" (van Gelder 1995, p. 353). In fact, as Bechtel notes, there is nothing about our ordinary understanding of representation that would preclude them from standing in a number of very complicated relations to processes and mechanisms that they influence. For example, something functioning as an internal map might be modified by various feedback mechanisms that are, in turn, subtly influenced by the behavior the map itself helps generate. We can't say something isn't a representation just because it participates in complex and subtle interactions. No, the reason the weighted arms aren't representations is not because their role is too nuanced; it is because their role is simply not representational in nature. They are functioning as a mechanism for causally linking engine speed to valve position, and not as any sort of representational stand-ins. Second, Bechtel is certainly correct in pointing out that the way he thinks about representation is shared by many contemporary theorists and researchers, especially in the areas of research associated with the computational neurosciences. As we noted in chapter 4, representation-qua-causal-mediation is basically *the* conception of representation at work in this very broad range of theories. The correspondence between Bechtel's analysis and these scientific theories is real, but it doesn't show his analysis is correct or that it yields an explanatorily valuable notion of representation. Quite the contrary, given the difficulties with his analysis that we've just gone over, it shows that there is something seriously wrong with the way representation is conceived of in these accounts.

Bechtel is not the only philosopher to respond to DST by offering an analysis of representation that attempts to show that simple devices like the Watt governor actually are representational systems after all. Another defense of representationalism adopting this tack is offered by Anthony Chemero (2000). Chemero begins by presenting the following definition of representation, which he regards as relatively restrictive:

A feature R_o of a system S will be counted as a *Representation* for S if and only if:
(R1) R_o stands between a representation producer P and a representation consumer C that have been standardized to fit one another.
(R2) R_o has as its proper function to adapt the representation consumer C to some aspect A_o of the environment, in particular by leading S to behave appropriately with respect to A_o, even when A_o is not the case.
(R3) There are (in addition to R_o) transformations of R_o, $R_1 \ldots R_n$, that have as their function to adapt the representation consumer C to corresponding transformations of A_o, $A_1 \ldots A_n$.

Because Chemero's definition of representation appeals to representational "producers" and "consumers," it initially appears that his definition is circular in that representational concepts show up in the definiens. However, it soon becomes clear that, like Bechtel, Chemero does not mean anything fancy by these notions. For Chemero a representation consumer is nothing more than some "down-stream" component or process that is influenced by the alleged representation in some systematic and goal-oriented way. Thus, once again, the weighted arms in the Watt governor qualify as representations and the shut-off valve qualifies as a representation consumer. As with Bechtel, this leads Chemero to adopt the intentional stance toward the governor:

It is the function of particular arm angles to change the state of the valve (the representation consumer), and so adapt it to the need to speed up or slow down. For consider that the governor was *designed* so that the arm angle would play this role; that is, arm angle tokens are part of the governor *because* they lead to appropriate control of the engine speed. So the function of the arm angles is to control the speed of the engine, and since each angle indicates both a speed and the appropriate response to that speed – is both a map and a controller – it is an action oriented representation, standing for the current need to increase or decrease the speed . . . Furthermore, the arm angle can 'be fooled', causing behavior for a non-actual engine speed; imagine what would happen if we used a flat surface to hold the arm at an artificially high angle . . . Thus, the arm angles of the Watt Governor are action-oriented representations. (Chemero, 2000, pp. 632–633)

So once again something serves as a representation in this sense when by design it serves to causally influence a process in response to *being* influenced by some other condition. While Chemero admits that such a notion may do no real explanatory work, especially in explaining simple systems like the Watt governor, he is nevertheless committed to the *metaphysical* position that representations exist whenever his criteria are met. Since I have already shown what is wrong with this position, it might be helpful to see just how *un*restrictive this account is by using the same logic and substituting an automobile engine for the Watt governor:

It is the functioning of particular spark plug firing rates to change the state of the drive shaft (the representation consumer), and so adapt it to the need to speed up or slow down. For consider that the engine was *designed* so that the spark plugs would play this role; that is, the spark plug firing tokens are part of the engine *because* they lead to appropriate control of the drive shaft speed. So the function of the spark plug firings is to control the speed of the drive-shaft, and since each firing rate indicates both the accelerator pedal position and the appropriate response to that accelerator pedal position – is both a map and a controller – it is an action-oriented representation, standing for the current need to increase or decrease the speed . . . Furthermore, the spark plugs can 'be fooled', causing behavior for a non-actual accelerator pedal position; imagine what would happen if we manipulated the accelerator cable leading to the car engine and thereby caused the spark plugs to fire at an artificially high rate . . . Thus, the spark plugs of the automobile engine are action-oriented representations.

The point here is that if you have an account of representation that entails that a spark plug is functioning as a representation , then you have a faulty *metaphysical* account of representation. The problem with Chemero's view is that he too confuses causal mediation with a representational role. Because of this, his account (like Bechtel's) fails to properly identify the demarcation line between representational and non-representational systems.

Part of my attack on these accounts of representation is to show that they are so weak that they lead to absurd consequences; for instance, that spark plugs are functioning as representations. But, of course, as with any intuition-based *reductio ad absurdum* argument, my critique will carry little or no weight against those who are willing to simply embrace the counter-intuitive consequences of their theory. After all, if you are comfortable claiming that the weighted arms of the Watt governor serve as representations, then you probably won't be bothered by the idea that spark plugs are representations as well. So as with the tacit notion, it is appropriate to reflect on who has the burden of proof. *Prima facie*, things that function as mere causal mediators don't come close to what we ordinarily think of as representational states. Hence, the burden is upon those who want to suggest this radically different way of thinking about representation is appropriate and beneficial. We need to be told why we should dramatically expand our conception of representation so that it now applies to causal relays. Merely redescribing basic causal processes in terms of "information carrying" is not sufficient. This is analogous to demonstrating that robots can actually feel pain by simply defining pain as any internal state that generates avoidance behavior. We are properly skeptical about dramatically deflationary definitions when they are offered to explain qualitative mental states. We should be equally skeptical when such a strategy is employed to account for representational states.

It is perhaps fair to insist that I have a burden as well. I am claiming that a very popular way of thinking about representation is, in fact, completely out of sync with our ordinary, intuitive understanding of representation. But if it is out of sync in this way, then why is the receptor notion so popular? Don't I need to offer some sort of account of this apparent contradiction in the way people think about representation? This a fair challenge, but one that I believe is easily met. For starters, it is not at all uncommon for our surface attitudes and practices to conflict with our more fundamental conceptual convictions. One role of philosophy is to make evident how our deeper conception of some phenomenon clashes with a popular perspective.[3] As we saw in chapter 4, and in our discussion

[3] For example, Singer (1972) and Unger (1996) have convincingly argued that our superficial attitudes toward people in desperate circumstances contradicts more deeply held moral principles and convictions. While we act as though we have no moral obligation to help those suffering from lack of food or medical care, a number of hypothetical cases suggest that deep down, we actually are committed to moral principles that entail such an obligation.

of mindless representation in this chapter, it is plausible to assume that the receptor notion in cognitive science is derived from our use of receptor-like structures that exist in the external world. *We* use things that reliably respond to something else to make accurate inferences about various things in the world. The rising mercury in a thermometer literally informs us that the temperature is rising. Thus, we have grown accustomed to treating things whose states nomically depend upon other conditions as having a representational role simply because they have such a role in certain (i.e., mindful) contexts. The error occurs when we overlook the fact that when the cognitive agent is removed from the picture and replaced by something much less sophisticated, a significant change happens in the nature of the process. We neglect to notice that the process becomes just like other causal processes that we intuitively judged to be *non*-representational in nature. Hence, we can account for the popularity of this perspective on representation without supposing it accords with our deeper understanding of representational function.

One final defense of representationalism worth examining is presented in Clark and Toribio (1994) and, to some extent, later in Clark (1997). Like Bechtel and Chemero, Clark and Toribio complain that, at best, the dynamicist's arguments apply only to the explicit symbolic representations associated with CCTC, and not the more modest notions of representation that appear in theories like connectionism. They argue that the dynamicist's case against representationalism falls short because it ignores what they call the more "representation-hungry" problem domains. Representation-hungry cognitive tasks require either "reasoning about absent, non-existent or counter-factual states of affairs" and/or that, "the agent to be selectively sensitive to parameters whose ambient physical manifestations are complex and unruly (for example, open-endedly disjunctive)" (Clark and Toribio 1994, p. 419).

The first condition is problematic since trivially, if the problem-solving strategy can be characterized as "*reasoning about*" something (be it counterfactual states of affairs or not), then, of course, the system is using states that represent. This is a bit like saying that representations are required in cognitive tasks that are representational in nature. The second condition, however, is more interesting. The suggestion is that whenever a system responds selectively to different features or combinations of features in the input, the system is using representational states. As they put it,

> the cognizer must be able to treat inputs whose immediate codings (at the sensory peripheries) are quite similar as deeply different (dilation) and conversely, to treat inputs whose immediate codings are quite different as deeply similar (compression).

On the modest reading which we recommend, internal states developed to serve this end just are internal representations whose contents concern the states of affairs thus isolated ... Any process in which a physically defined input space is thus transformed so as to suppress some commonalities and highlight others is, we claim, an instance of modest representation. (Clark and Toribio 1994, pp. 421, 427)

So what is thought to make an internal state into a real, though modest, representational state is its capacity to handle complex forms of input to the system. This comes in two forms – responding differently to similar sorts of input, and responding in a similar manner to very divergent inputs. Let's consider each of these situations more carefully.

In the case of internal states that respond differently to superficially similar inputs – what they call input "dilation" – it is difficult to see, on closer inspection, just how this is supposed to be significantly different from a case of straightforward causation, albeit causation that involves hidden factors. Unless the system is responding in a random manner, there must be some type of difference between the superficially similar inputs that gives rise to dissimilar responses, even if *we* aren't sensitive to those differences. If there are internal states that respond "selectively" to these hidden features, then they are reliable responders to those features. If they are incorporated into the processing because of this response capacity, then they serve as causal mediators that are perhaps highly fine-tuned in their responsiveness. But nothing about this suggests the presence of representations. Note that the weighted arms of the Watt governor respond quite dramatically to subtle changes in the engine speed – changes that might be completely invisible to us. But that doesn't mean the arms are serving as representations.

In the case of grouping disjunctive inputs – what Clark and Toribio call the "compression" of input – the crucial assumption appears to be that something cannot function as a mere causal mediator or relay circuit (i.e., function in a non-representational manner) if it is designed to respond to different triggering conditions. But this assumption seems quite implausible. A relay circuit is still a relay circuit, even if the triggering causes that "toggle the switch," so to speak, are multi-faceted or disjunctive. In fact, this is how many actual relay circuits are designed to work – bridging a causal gap between a disjunctive array of triggering conditions and whatever process or processes that are supposed to result from those conditions. In some cars the firing rate of the spark plugs is determined *either* by the position of the accelerator *or* the setting of a cruise control system. Yet this disjunctiveness doesn't make the spark plug a representation. Indeed, since no two input conditions are exactly the same, the class of triggering conditions for any causal process is always in some way disjunctive.

On a related point, Clark and Toribio emphasize the need for internal representations whenever a system responds to more abstract properties, such as things that are valuable. Of course, if a cognitive system is using a "valuable things" concept to do this, then the system is using some form of representation. But the important question is how do you get a state that is functioning in a physical system as a "valuable things" representation? The answer they offer is by being a functional state that reliably responds to all and only valuable things. Invoking such a state is, they insist, "just to invoke the idea of an internal representation" (1994, p. 420). But invoking a state that reliably responds to an abstract kind (and being designed to do so) is not to invoke a representation of that kind, any more than invoking a state that responds to a straightforward physical kind makes something a representation of that physical kind. The set of valuable things is no doubt multiply realized in different physical kinds. Hence, any structure that is activated by valuable things will need to be activated by a disjunctive set of physical conditions. Yet as we just saw, disjunctiveness in the triggering conditions does not give rise to representation. To repeat (yet again), something serving as a reliable respondent is not serving as a modest type of representation, even if it is a respondent to disjunctive states of affairs.

Thus, Clark and Toribio's attempt to supplement mere internal states to show how modest representations could come about doesn't actually add anything recognizably representational to the picture. What they present is a more sophisticated sort of relay system, to be sure, but not a set of conditions in which anything serves to stand for, or stand *in* for, anything else. In their analysis, they suggest that the burden is upon critics to show that an internal state playing this role is not a representation.[4] This is certainly possible, as I've just shown. But this also gets the explanatory burden backwards. The burden is upon those who posit representational states as part of their explanatory apparatus to defend this characterization of internal states. For those who invite us to regard structures as representations even though they function just like things we *wouldn't* normally consider representational in nature, the burden is upon them to explain why this dramatic shift in perspective is warranted. As far as I can see, this burden has not been met.

I have focused upon these philosophical defenses of representationalism for a number of reasons. First, a good part of my own analysis of

[4] They state: "What the committed anti-representationalist thus really needs to do is to convince us that conformity to the modest notion just rehearsed is really not enough to establish the truth even of the general representationalist vision" (1994, pp. 415–416).

representational notions is in direct conflict with the accounts offered here. To some degree, all of these authors are attempting to develop and promote versions of the receptor notion. Because I've argued the receptor notion doesn't work, it is essential for my account that we consider ways in which someone might try to make it work. Seeing how these attempts fail provides further support for my claim that the receptor notion doesn't really belong in cognitive science. Second, as all of these authors argue, the conception of representation they defend can be found in a wide range of theories of cognition, especially theories in the computational neurosciences. Indeed, the prevalence of this way of thinking about representation in the brain sciences is often assumed to demonstrate its explanatory value. Yet, as I've argued above, the explanatory utility of a representational posit is not really dependent upon its popularity. Instead, it depends on the way it actually allows us to understand how cognition involves the employment of an internal state that serves to stand for (or stand in for) something else. Since this particular conception of representation fails to do this, its popularity reveals that there is a deep and widespread misconception about the explanatory apparatus employed in several cognitive theories. While I have no real interest in endorsing the dynamic (or any) account of cognitive processes, its challenge to representationalism cannot be answered by invoking non-representational states and calling them representations. While it may be true that the dynamicists go too far in their sweeping denial of *any* need for representations, they are right to reject the more modest forms that involve simple causal mediation.

Finally, despite the problems with these attempts to rescue a notion of representation, these writers are to be commended for exploring what I think is an important but under-appreciated issue, namely, how it is that something can actually function as a representation in a physical system. The difficulties associated with these efforts reveal just how tough it is to accommodate both our fundamental understanding of representation, *and* our desire to provide a naturalistic account of cognition. Given the central importance of representation throughout cognitive science, it is no small embarrassment that we are still unclear on how representations are supposed to operate, as such, in a biological or mechanical system. This is arguably one of the most pressing concerns in the brain sciences – more so than the related (but different) matter of explaining representational content in naturalistic terms. Hence, a close look at what doesn't work helps to give us a better sense of what might work. An idea initially embraced by most of these writers is that representations are things that "stand in" for something else. I believe this is the right way to look at things. Their mistake comes from contaminating this idea with a very different way of thinking about representation – one that treats representations as things that serve to respond to other things.

6.3 IMPLICATIONS OF A NON-REPRESENTATIONAL PSYCHOLOGY

In this final section, I want to consider some of the meta-scientific and metaphysical implications of the arguments and analyses presented in earlier chapters. Nothing I have said so far is intended to have any *direct* consequences about the way the mind-brain works. That is an empirical matter, and it should be decided through empirical research and theory testing. Ultimately it is to be decided by determining which theories do the best job of accommodating the diverse range of data. Instead, my analysis is about the nature of psychological theories themselves, and the sorts of things they posit as elements of the mind/brain in their attempts to explain cognition. Much of my analysis has been intended to show that there is a profound misunderstanding about one of those posits, cognitive representations, and that this has led to the wide-spread invoking of so-called representational states by theorists who, in reality, neither need nor use *actual* representational states in their proposed explanatory frameworks. If, as Kuhn suggests, paradigms exert considerable influence over what investigators see and think, then the representational paradigm has prevented people from appreciating just how far contemporary research has actually drifted away from invoking states that serve to represent. In section 6.3.1, I want to consider some of the things this might mean for the so-called "cognitive revolution" and future of the brain sciences.

Second, while my analysis does not involve direct consequences for how the mind-brain works, it does of course entail *indirect* consequences. If I'm right, and *if* one of these covertly non-representational theories should prove correct, then it will turn out that the brain does not actually use internal representations when performing various cognitive tasks. This is an outcome that has profound consequences not just for the brain sciences, but also for our ordinary, commonsense understanding of minds and of each other. Since on most accounts, the accuracy of commonsense psychology depends upon the existence of internal representational states like beliefs and desires, the non-existence of such states would demolish folk psychology, an outcome Jerry Fodor plausibly describes as, "beyond comparison, the greatest intellectual catastrophe in the history of our species" (Fodor 1987, p. xii). So while these issues and debates might seem like quaint intellectual disputes, quite a lot rides on these matters regarding the way we think of ourselves. It will therefore pay to reflect a bit more on these matters – something I will do in section 6.3.2.

6.3.1 A revolution in reverse

Let's assume that my arguments are sound and that my analysis of these different notions of representation in cognitive science is correct. That would mean that two very popular and prevalent ways of thinking about representations in cognitive science – the receptor notion and the tacit notion – are fatally flawed, and that a wide range of theories that claim to be representational in character, like many theories in the computational neurosciences, actually aren't. It would also mean that theories that appeal to internal input and output representations and S-representational notions, like many CCTC theories, are using legitimate and explanatorily valuable concepts of representation. Insofar as the former range of theories are often thought of as successors to the latter, what would all of this imply about the direction of cognitive science research?

One clear implication is that the cognitive revolution is, in this one very important respect, moving backwards. There are many elements of cognitivism that demarcate it from behaviorism, including an acceptance of conscious phenomena and a more sophisticated understanding of psychological mechanisms. But certainly one of the most distinguishing (and to some degree defining) hallmarks of cognitivism has been a strong commitment to internal representations as an explanatory posit. This assumption is so deeply ingrained that, as we've seen, some cognitivists consider it folly to even question the explanatory value of representational posits (Roitblat, 1983). And yet, as we've also seen, this commitment to representational posits has been unwittingly and surreptitiously abandoned by those working in a significant and growing line of research. For theories invoking the receptor and/or tacit notions to account for cognitive capacities, these capacities are actually being explained by hypothesized operations that have nothing to do with internal representational states, processes, roles, mechanisms or the like. These theories are something of a reversion, albeit an inadvertent reversion, back to the earlier behaviorist strategies of explaining the mind without positing internal representations. Of course, not all of the newer non-classical theories invoke dubious conceptions of representation. As noted in chapter 3, both Grush (1997, 2004) and Ryder (2004) attempt to develop an S-representational theory for neural networks. But these theories are somewhat exceptional. The more typical non-classical theories appeal to internal structures and processes that are described as playing a representational role, when in fact they are playing no such role. In fact, the role they are actually playing is not so far removed from what at least some thought was taking place in the head, before the advent of cognitivism.

We can see this better by comparing the functional role of the pseudo-representations with some of the views about internal processes held by those traditionally associated with the behaviorist tradition, including not just neo-behaviorists, like Tolman (who did invoke representation-like states), but also those who deliberately avoided positing anything thought to have a representational role. Receptor representations function as causal mediators between stimuli and internal and external responses, and tacit representations are really just the dispositional properties of the functional architecture itself. Along similar lines, many straightforward behaviorists, though typically wary of intervening variables, recognized the need for internal states with the job of causally mediating between different forms of stimuli and responses. After all, no behaviorist thought the causal bridge between stimuli and response was magic. Watson (1930) at least partly recognized the role of the nervous system in forging connections between perceptual inputs and behavior, and Skinner (1976) characterized internal responses and internal variables as "private events" that play an important role in responding to specific stimuli. While these and other radical versions of behaviorism allowed for a limited explanatory role for hypothesized intervening variables, other versions of behaviorism were committed to more robust "mediator" states and structures that, in terms of their functionality, are not radically different than receptor or tacit representations (Homme 1965; Kendler 1971; Osgood 1956).

One such mediative version of behaviorism was developed in the 1930s by Clark Hull, who recognized the inadequacy of trying to explain various types of learning as a straightforward S-R reflex (Hull 1930, 1931, 1932; Amsel and Rashotte 1984). Using a variety of conventional behaviorist learning principles (many borrowed from Pavlov), Hull developed a sophisticated account that invoked internal or "covert" stimuli-response events that could be chained together through various feedback associations. These response chains typically included internal mediators with the functional role of responding to specific types of stimuli and generating some form of appropriate response. For instance, Hull postulated that each overt response to an external stimulus would itself generate an internal proprioceptive stimulus that could serve to trigger further behavioral responses. If this sequence was beneficial to the organism, it would become reinforced, making it possible for a single external stimulus to generate a long and elaborate response chain. Hull claimed that this wiring process accounted for much of the flexibility and sophistication seen in animal behavior, including the capacity to anticipate various environmental conditions. In his account, these internal stimuli function as reliable mediators between different forms of environmental conditions and particular goal or anticipatory behaviors. As Amsel and Rashotte note, "Hull's proposal for extending S-R concepts to complex behaviors involved the identification of an array of unobservable stimuli that could enter into associative relations with responses and thereby, along with observable external stimuli, exert control over behavior . . . For him, an animal's knowledge of its world is, to use a modern term, 'encoded' in response chains" (Amsel and Rashotte 1984, p. 35).

While Hull does describe these internal response chains as constituting the animal's "knowledge," he never suggests that the internal stimuli play any sort of representational role or should be treated as a type of representation. Nor does he claim that they serve to relay informational content about the external stimuli or the activities that trigger them. As Baars notes, "Hull insists that his $r_g - s_g$ associations are *not* representations – they are merely physical stimuli and responses that become associatively conditioned . . ." (Baars 1986, p. 58). For Hull, the role of his internal mediating structures was purely causal in nature, though he claimed they gave rise to complex and sophisticated sequences of behavior. Yet it is difficult to see any deep difference between the functional role of Hull's internal mediators and the various forms of receptor representations that have become prevalent in so many contemporary cognitive theories. The same goes for many other non-representational intervening states hypothesized by many other behaviorists. In both paradigms, you have the positing of internal mechanisms with the job of regularly and reliably bringing about certain responses when, and only when, certain conditions obtain. In terms of details, of course, there are many differences. But in terms of the primary job ascribed to these internal intervening states, we see the same basic set of conditions.

A similar point applies to the notion of tacit representation. In fact, virtually all behaviorists acknowledged that learning brings about significant changes in the functional architecture of an organism's nervous system. Moreover, they recognized that these changes could be characterized as the acquisition of various dispositional properties (Skinner 1976). What they deliberately avoided, however, was describing these changes as the attainment of representational states, even of a tacit form. Of course, modern accounts of learning provided by such processes as back-propagation are far more sophisticated than anything proposed by most behaviorists. But the sort of tuning of connection weights seen in most connectionist accounts of learning, brought about by numerous trials, is something most behaviorists would recognize as a friendly extension of their own non-representational stories of learning.

If the behaviorist's non-representational mediational states are really no different than the cognitivist's receptor and tacit representations, what should we conclude? One possibility is that the behaviorists were just confused, and failed to realize that their internal posits were actually playing a representational role. According to this perspective, these behaviorists were covert representationalists. But I would hope that the arguments and analysis presented in earlier chapters would rule out this interpretation, pointing instead in the opposite direction. The correct conclusion is that the contemporary cognitive theorists who appeal to receptor states and tacit representations are really not that different from many earlier behaviorists concerning the functional role of these central explanatory posits. To be sure, connectionists, cognitive neurologists, computational neuroscientists and the like, all offer theories that are far more explanatorily robust and predictive than behaviorists like Hull. The claim here is not that the new theories in the cognitive neurosciences are the same as, or no better than, the earlier behaviorist theories. In fact, today we see a level of sophistication regarding the internal machinery of the mind that goes way beyond what behaviorism had to offer, and I see no reason to assume these contemporary theories deserve the same fate as behaviorist accounts. Rather, my claim is that with regard to this one very important aspect – the positing of internal representations – these modern theories are far more like many behaviorist accounts than generally appreciated. Insofar as cognitive science is moving in a direction that embraces these connectionist-style explanatory frameworks, then we are seeing a regression back to theories of the mind and mentality in which cognitive representations play no real explanatory role. The cognitive revolution, in this one very important respect, would be moving in reverse, returning to a non-representational psychology.

The positing of representations has traditionally served to demarcate more than just the difference between cognitive and behaviorist theories of the mind. Representational posits are also often used to determine the appropriate explanatory level at which a given theory belongs. We saw in chapter 2 that Marr provided a three-tiered template for dividing the levels of analysis of cognitive research. The top level provides a specification of the capacity or task being explained, the middle level provides the algorithmic explanation of the task, and the bottom level informs us of how all of this is implemented in the actual neurological wet-ware. In recent years, many have come to question this simple three-level picture (see, for example, P. S. Churchland 1986). A common complaint is that there are many more levels of organization and analysis than the three Marr suggested, and the boundaries between levels are far more fuzzy than is often assumed. As theories in the cognitive (or computational) neurosciences become more and more "low-level" and biological in nature, it is increasingly unclear why they should be regarded as truly psychological theories, as opposed to "merely implementational"[5] accounts of neurological mechanisms. What exactly is the difference between explaining some process by appealing to a lower-level, but nevertheless *cognitive* mechanism, and explaining the same process by appealing to a somewhat abstract characterization of an implementation-level neurobiological process?

One of the most common answers offered is that at the cognitive level, theories appeal to representations while at the purely physiological level, the language of representations is not employed. For example, in explaining what it is about neuronal processes that make them computational, Churchland, Koch and Sejnowski tell us we should "class certain functions in the brain as computational because nervous systems *represent*" (Churchland, Koch and Sejnowski 1990, p. 48). Thus, theorists who want to be seen as working in the field of the cognitive or computational neurosciences (as opposed to, presumably, just the neurosciences) have offered accounts of mechanisms or processes that are described as representational in nature. But because these theories often rely upon the receptor and tacit notions of representation, these accounts of the nervous systems (including those endorsed by Churchland and Sejnowski 1992) do not involve states that are *actually* playing a representational role. If this manner of demarcating computational and neuronal processes is correct, then many accounts often described as "computational" or "cognitive" in nature would actually fail to qualify. This is not to say that some other method couldn't be used to classify "higher" levels of analysis. Perhaps some sort of story about the level of organization, or perhaps the nature of the explananda, could serve a similar meta-theoretical role. Yet I'm inclined to think that our ordinary perspective on levels of analysis in psychology will simply be dropped. The traditional view that there is a privileged level at which the theoretical apparatus is uniquely suited to

[5] The "merely" is often intended pejoratively, as suggested in Fodor and Pylyshyn's well-known critique of connectionism (Fodor and Pylyshyn 1988).

psychological or cognitive explanations (while lower levels provide only the neuronal details of how a proposed model is implemented), is perhaps a perspective ill-suited for understanding the mind. We don't treat biological processes as requiring a uniquely "organic" level of analysis, with physical descriptions of intracellular or enzyme processes treated as merely providing implementational details. Physical processes inside living tissue, and therefore theories explaining those physical processes, are regarded as organic (for the most part) all the way down to the molecular level. Perhaps once we abandon the view that cognitive explanations must always invoke representational posits, we can also adopt a broader-ranging understanding of what counts as cognitive processes and therefore a less restrictive understanding of what counts as a cognitive level of analysis.

If I'm right that there has been a radical about-face in the direction of cognitive research, then why have so few people noticed the conversion? There are, I believe, several possible explanations, some we have already discussed. Representational notions are theoretically tricky, philosophically complex, and generally quite confusing. Many of the misunderstandings I have tried to expose required considerable philosophical analysis to expose. It is not surprising that, for example, people have come to believe that because an internal neurological state is nomically dependent upon some external condition, it is therefore a representation of that condition. It is not surprising because *we* sometimes use non-mental things with nomic dependency properties as representations, and it is easy to overlook the differences between this sort of set-up, and the set of conditions existing inside a brain. So when people propose neurological representations of this sort, it very difficult to see that, in truth, no representational function is actually implemented.

But I also believe there is a sociological, or meta-scientific explanation for the general tendency to proliferate representational posits even when they do no real explanatory work. As with any paradigm, there are assumptions and prejudices built into cognitivism that are a reaction to an earlier framework. Certainly one such assumption is the idea that behaviorism failed, at least in part, because it tried to explain the mind without acknowledging inner mental states, especially cognitive representations. Thus, for many contemporary theorists, an essential element of any serious theory about cognitive processes is an appeal to internal representations. If a theory fails to do this, then it certainly runs the risk of being tarnished with the behaviorist stigma. As I've noted in earlier works (Ramsey 1997), this anti-behaviorist bias can serve as a strong motivation for calling a posited state a representation, even when the functional role actually assigned to it is not, in any real sense, representational in nature. Since Kuhn (1962), we have come to recognize that investigators working within a given theoretical paradigm often interpret results and make observations in a manner that is deeply influenced by their theoretical assumptions and commitments – something Kuhn called "theory-laden perception." According to Kuhn, the effect can be so strong that researchers come to see things that aren't really there, or at least see structures as doing one thing when they are actually doing something quite different. It is not implausible to suppose that cognitive scientists and philosophers who view themselves as participating in post-behaviorist theorizing are especially prone to treating internal structures as representations, even when they are playing no such role.

One further reason I believe people have been blind to the non-representational turn in cognitive research is because our ordinary conception of mentality makes it practically unthinkable that there could be an account of mental activity that doesn't involve representational states. For example, the most natural and intuitive way to treat our five senses is to treat them as faculties that serve to produce representations of the "outer" world that make up our "inner" experience. Thus, eyes are easily regarded as something like video cameras, serving up images that are then displayed somewhere in the brain for our inner "viewing." Dennett (1991b) has called this mythical place where representations are presented for conscious experience the "Cartesian Theater." Dennett's main point is to reveal a common misconception about consciousness. But his Cartesian Theater also suggests something about the centrality of representation to our commonsense understanding of mental phenomena. If the function of sensory systems is to produce something as output, then the most natural way to view that output is to assume that it comprises representations that are (of course) re-presented for some internal viewer. Even if, upon reflection, we realize there can be no inner viewer (or hearer, smeller, etc.), it is extremely difficult to abandon a general understanding of the mind that presupposes inner states functioning to somehow represent the world to "ourselves." Given that folk psychology itself is fully committed to the existence of mental representations like beliefs and desires, our folk psychology makes it very difficult to see scientific psychology as moving in a non-representational direction, even when it actually is.

But what if cognitive science not only moves in a non-representational direction, but actually succeeds in producing a theory of cognition that is both non-representational and, it turns out, extremely successful in accounting for cognitive capacities. What if some non-representational version of a theory in the cognitive neurosciences should ultimately prove correct? What would this suggest about our ordinary conception of the mind?

6.3.2 Eliminativism revisited

As I noted in prior chapters, in an earlier paper I co-authored with Stephen Stich and Joseph Garon we argued that if a family of connectionist models of memory and inference should turn out to provide a true theoretical framework for understanding these psychological processes, then such a result would entail eliminative materialism (i.e., the claim that there are no such things as beliefs and other propositional attitudes). Connectionist models of memory threaten eliminativism because their long-standing representations are tacit and superpositional, and therefore lacked the type of functional discreteness that we claimed beliefs and other propositional attitudes require. As I stated in the last chapter, I now think a far more direct argument is possible. Many connectionist models are incompatible with folk psychology not because they employ representational posits of the wrong sort, but because they employ no representational posits, period.

While both my earlier and current endorsements of eliminativism are only conditional, based on what would follow *if* a certain brand of connectionism should prove correct, I have often found myself pressed to defend the possibility and even the intelligibility of eliminativism on various occasions (see, for example, Ramsey 1991, 2003). While I do regard eliminativism as coherent, I must also confess that I have considerable sympathy with those who regard it as insanely implausible. It seems quite difficult to take seriously the idea that the beliefs and desires that so obviously make up a large part of my conscious experience don't actually exist. In fact, the introspective evidence we have for their reality seems substantially better than any possible data that could be used to support a theory of cognition that denies their existence. We can, of course, challenge the veracity of introspective access to our own mental states. However, to be able to use any empirical data, we presumably need to be confident that the data actually exist, and it is hard to see how we can be confident about that without also being confident that we possess accurate mental representations about the data. So in promoting a non-representational analysis of a wide range of cognitive theories, one question I often return to is this: Seriously, how could there be a true theory of the mind that leaves out the representational states that are so obviously a part of our mental lives?

Well, here is one (admittedly quite speculative) way it might happen. No eliminativist denies the existence of public utterances and statements.

No one denies that we talk. Moreover, no one denies that we sometimes talk to ourselves. Now as some authors have suggested (Jaynes 1976; Dennett 1991b), it is possible that the most vivid, consciously accessible thoughts that so apparently constitute an important aspect of our mentality are actually a type of sub-vocalized private talk. These internal utterances perhaps stem from the same cognitive operations that give rise to public utterances; it's just that in the sub-vocalization case, we leave out the final speech-producing apparatus (the larynx, the mouth, etc.). How any aspect of this process becomes part of our conscious awareness is, like anything else involving consciousness, an utter mystery. But let's suppose that something close to the end state of the utterance-producing processes (but prior to actual vocalized statements) is what we experience as our own very conspicuous and seemingly undeniable beliefs and desires. We could then have two cognitive levels – one involving a basic non-representational architecture, and another, perhaps newer and more peripheral representational system that is closely connected to our capacity to utter sentences. A similar proposal for such a split-level system has been offered by Keith Frankish (2004). Frankish proposes that our cognitive system is made up of both a basic mind, which is non-conscious and in certain respects non-linguistic, and a "supermind," which gives rise to flat-out, language-like conscious thoughts. On Frankish's account, supermind thoughts play the sorts of causal roles associated with folk mental representations, and thus, even though the deep structure of the mind might be non-representational in nature, cognitive explanations of behavior would still need to invoke causally salient representational states. In this picture, no accurate theory about the processes that give rise to behavior and reasoning could avoid positing belief-like states, and thus, folk psychology would be vindicated.

But now suppose this last feature of Frankish's two-tiered account – that conscious thoughts are causally salient – turns out to be wrong. Suppose the best accounts of cognition are like the ones I've described as non-representational in nature and further suppose that our sub-vocalized occurrent thoughts are merely *output* arising from deeper processes. As output, these sub-vocalized utterances play no further role in the production of behavior or significant cognitive operations that underlie our various psychological capacities. Perhaps they would be relevant to verbal behavior, but let's assume that they don't do much. If this were so, we would have the means to offer what might best be described as a sane version of eliminativism. We could acknowledge and accommodate our most vivid evidence supporting belief-like states, and at the same time accept that commonsense psychology is severely flawed and that the inner mechanisms and processes of the mind are non-representational in nature.

How would this work? The key would be to accept the reality of what seems so indisputable – consciously experienced representational states, i.e., thoughts – while at the same time leaving open the possibility of true theories about cognitive processes and capacities that are non-representational in nature. On the sub-vocalization hypothesis, conscious thoughts are the *effects* of more central psychological operations and processes. It would be these central psychological processes that non-representational theories, like the family of connectionist models mentioned earlier, would actually serve to explain. If these theories are true, then the bulk of our cognitive machinery and processes don't involve states functioning as representations. If, let's say, we want to know how we generate grammatical sentences, or recognize faces, or decide what our next move should be in chess, then, ex hypothesis, the theory that actually explains the internal workings of the cognitive apparatus responsible for these capacities would not posit any internal representations. Would this be surprising? Certainly – it would go directly against the conventional wisdom of both scientific and commonsense psychology. But would this be insanely implausible? Given that our introspective access to these internal operations and processes is now recognized as often limited and misleading (see, for example, Nisbett and Wilson 1977; White 1988) and given that even commonsense psychology allows that we are often blind to the deep inner workings of our own minds, this would not be terribly farfetched. It would not be utterly ridiculous to learn, say, that the process *giving rise to* a sub-vocalized expression with the content "Because my opponent's queen is in position X, I should try to trap it by moving my knight to position Y" is itself a process that uses no internal states functioning as representations. There would still be the conscious output that would somehow represent the chess game – output that we can be confident exists. But the actual processes giving rise to both my moving my knight and my telling myself (sub-vocally) that I ought to move my knight could itself be representationless.

So in this sort of scenario, the states it seems so crazy to claim are unreal – occurrent, consciously accessible thoughts – would not be denied to exist. They would be accepted as a part of our mental lives and perhaps, at some level, would play a role in coordinating our linguistic behavior. Yet at the same time, there would be no inner representational states playing the type of causal or inferential roles that folk psychology ascribes to beliefs and desires. The cognitive machinery where the real action takes place would have nothing that possesses the central feature associated with propositional attitudes. If we ask whether folk-psychological terms like "belief" pick out anything real, we would therefore have something of a mixed verdict. On the one hand, it could be argued that much of the time we use such terms to refer to occurrent conscious thoughts. Since these states would indeed exist in the form of sub-vocalized expressions, then, in this sense, the posits of folk psychology would successfully denote something real. On the other hand, because, ex hypothesis, these active thoughts would be output of central psychological processes, they would fail to possess the sort of causal properties normally associated with beliefs and other propositional attitudes. The cognitive architecture that would generate behavior and give rise to various capacities would not involve representations. Insofar as folk psychological explanations (and predictions) appeal to states defined as having both representational and causal properties (states like beliefs), it would turn out that these explanations and predictions appeal to states that do not exist. In this sense, there would be no such things as beliefs.

I suggest this as only one speculative story about how it might happen that non-representational theories about the cognitive architecture and processes could turn out to be true. If the central claims of this book are correct, then a great deal of cognitive research today involves models of the mind that are indeed non-representational in nature. Given that non-representational psychology is eliminative psychology (since beliefs and folk mental representations are essentially representational), it is at least mildly reassuring to know that it's possible for everything to turn out all right – sort of. In my crude account, the mind would still involve the sorts of states nearest and dearest to our experienced mental lives, even though these states would not play a large role in our best explanations of most cognitive capacities (except, perhaps, as a sort of output). It's possible, in other words, for non-representational psychology to come to the forefront of psychology without *completely* crazy results.

6.4 CONCLUDING COMMENTS

The rise of cognitivism in psychology is often regarded as nothing short of a Kuhnian revolution, with cognitive representations playing the leading role in defining the new paradigm. I've suggested here that there is now something like a counter-revolution taking place, even though most of the participants don't realize it. This unawareness is due to a serious deficiency in our understanding of the kind of functional roles representational states do and don't play, and thus in the kind of explanatory role representational posits do and don't play in cognitive theories. There are two conceptions of representation that are appropriate and generally found in the classical computational framework – the IO notion and the S-representation notion. However, these are being gradually replaced by two different conceptions – the receptor notion and the tacit notion. The problem is that the latter two conceptions, on closer inspection, don't actually involve anything playing a recognizably representational role. Of course, the S-representation notions and IO notions have their problems too, and more work is needed to generate a full theory of each. But at least with these notions we have a way of thinking about representation that reveals how something can function as a representation in a physical system. I don't see this with the receptor or tacit notions. We can use graphite to make synthetic diamonds, even though there are problems that must be overcome. But we can't use lead to make gold. As I see things, the kind of difficulties that confront the S-representation and IO notions are of the former type, whereas the problems that confront the receptor and tacit notions are of the latter sort. Contrary to conventional wisdom, I believe these latter ways of understanding cognitive representation are unpromising non-starters. Consequently, while the newer non-classical models of the mind are perhaps exciting, promising, well-supported by the data and explanatorily fruitful, they are not, despite all that, *representational* theories of the mind.

This has been the central theme of this book. Yet I hope the reader will take away several other lessons from my analysis. A more complete listing would include these central points:

(1) There is an urgent need for more work on what it means for something to function as a representation in a physical system. While teleology is regularly invoked in philosophical work on representation, it is normally used to handle problems associated with content, like the problem of error. It is important not to think that a theory of content that leans heavily on the idea that something (somehow) functions as a representation is itself a theory of what it is for something to function as a representation.

(2) The standard account of why the classical computational accounts of cognition need to invoke representational states is flawed, and places too much emphasis on our commonsense understanding of mental representations. In truth, the notions of representation that do explanatory work in the classical framework have little to do with commonsense psychology, and are instead built into the sorts of explanatory strategies classical theories offer.

(3) You can't vindicate commonsense psychology by identifying folk posits with states that don't function as representations, hoping that the identification itself will make the states more representation-like. Instead, you must first have theories that posit states that serve as representations in their own right, and then see if folk notions can map onto these.
(4) Of the two most widely invoked conceptions of representation invoked in cognitive science – the receptor notion (based on nomic dependency relations), and the S-representation notion (based on some kind of isomorphism) – only the second describes structures that actually function in a manner that is recognizably representational in nature. Receptor states do not play a representational role, and theorists should stop characterizing their job in such terms.
(5) Dispositional properties embodied in a cognitive system's functional architecture do not function as representations. Talk of tacit representations is really, at best, talk about non-representational dispositions.
(6) Given that many newer theories of cognition, like those found in the cognitive neurosciences, rely upon the receptor and tacit notions, these theories are actually non-representational in nature, despite the way they are normally characterized. Insofar as they typify a growing trend in cognitive science, the discipline is quietly and unwittingly moving away from representationalism. The cognitive revolution is, in this sense, moving backwards.

Scientific progress often requires looking at things in a radically different way than either tradition or intuition suggests. Both tradition and intuition tell us that the mind is a representational system – that if we are going to understand how it works, we need appeal to inner states that serve to stand for other things. It is time for us to carefully reconsider this outlook, a process I hope the arguments presented here will help initiate. My aim has not been to endorse a non-representational scientific psychology. Instead, it has been to establish just how much harder it is to actually *have* a successful representational science of the mind than is generally appreciated. It is too soon to tell how all of this will eventually turn out, or to tell whether the representational framework will survive. But the more open we are to different conceptions of the cognitive processes, including non-representational ones, the better our chances of actually getting it right.

References

Adams, F. and Aizawa, K. 1994. "Fodorian semantics," in S. Stich and T. Warfield (eds.), *Mental Representation: A Reader*. Oxford: Blackwell Publishing, pp. 223–242.

Amsel, A. and Rashotte, M. 1984. *Mechanisms of Adaptive Behavior: Clark L. Hull's Theoretical Papers, with Commentary*. New York: Columbia University Press.

Anderson, J. R. 1983. *The Architecture of Cognition*. Cambridge, MA: Harvard University Press.

2000. *Cognitive Psychology and its Implications: Fifth Edition*. New York: Worth Publishing.

Anscombe, G. E. M. 1957. *Intention*. Oxford: Basil Blackwell.

Baars, B. 1986. *The Cognitive Revolution in Psychology*. New York: Guilford Press.

Baker, L. 1987. *Saving Belief: A Critique of Physicalism*. Princeton: Princeton University Press.

Barlow, H. 1995. "The neuron doctrine in perception," in M. Gazzaniga (ed.), *The Cognitive Neuroscience*. Cambridge, MA: MIT Press, pp. 415–435.

Bechtel, W. 1998. "Representations and cognitive explanations," *Cognitive Science* 22: 295–318.

2001. "Representations: from neural systems to cognitive systems," in W. Bechtel, P. Mandik, J. Mundale, and R. Sufflebeam (eds.), *Philosophy and the Neurosciences*. Oxford: Blackwell Publishing, pp. 332–348.

Bechtel, W. and Abrahamsen, A. 2001. *Connectionism and the Mind: Parallel Processing, Dynamics and Evolution*. Oxford: Blackwell Publishing.

Beer, R. D. 1995. "A dynamic systems perspective on agent-environment interaction," *Artificial Intelligence* 72: 173–215.

Beer, R.D. and Gallagher, J. C. 1992. "Evolving dynamical neural networks for adaptive behavior," *Adaptive behavior* 1 (1): 91–122.

Bickle, J. 2003. *Philosophy and Neuroscience: A Ruthlessly Reductive Account*. New York: Kluwer/Springer Publishing.

Blakemore, R. and Frankel, R. 1981. "Magnetic navigation in bacteria," *Scientific American* 245(6): 58–67.

Block, N. 1986. "Advertisement for a semantics for psychology," *Midwest Studies in Philosophy* 10: 615–678.

1990. "The computer model of the mind," in D. Osherson and E. Smith (eds.), *An Invitation to Cognitive Science, Vol. 3: Thinking*. Cambridge, MA: MIT Press.

Boden, M. 1977. *Artificial Intelligence and Natural Man*. New York: Basic Books.
Brachman, R. and Levesque, H. 2004. *Knowledge Representation and Reasoning*. San Francisco: Morgan Kaufman.
Brooks, R. 1991. "Intelligence without representation," *Artificial Intelligence* 47: 139–159. Reprinted in Haugeland, 1997.
Chemero, A. 2000. "Anti-representationalism and the dynamical stance," *Philosophy of Science* 67(4): 625–647.
Churchland, P. M. 1981. "Eliminative materialism and the propositional attitudes," *Journal of Philosophy* 78: 67–90.
1989. *A Neurocomputational Perspective: The Nature of Mind and the Structure of Science*. Cambridge, MA: MIT Press.
Churchland, P. S. 1986. *Neurophilosophy*. Cambridge, MA: MIT Press.
Churchland, P. S., Koch, C., and Sejnowski, T. 1990. "What is computational neuroscience?" in E. Schwartz (ed.), *Computational Neuroscience*. Cambridge, MA: MIT Press.
Churchland, P. S. and Sejnowski, T. 1989. "Neural representation and neural computation," in L. Nadel, L. A. Cooper, P. Culicover, and R. Harnish (eds.), *Neural Connections, Mental Computation*. Cambridge, MA: MIT Press, pp. 15–48.
1992. *The Computational Brain*. Cambridge, MA: MIT Press.
Clapin, H. 2002. *The Philosophy of Mental Representation*. Oxford: Oxford University Press.
Clark, A. 1991. "In defense of explicit rules," in W. Ramsey, S. Stich, and D. Rumelhart (eds.), *Philosophy and Connectionist Theory*. Hillsdale, NJ: Lawrence Erlbaum.
1993. *Associative Engines*. Cambridge, MA: MIT Press.
1997. "The dynamical challenge," *Cognitive Science* 21 (4): 461–481.
2001. *Mindware*. Oxford: Oxford University Press.
Clark, A. and Toribio, J. 1994. "Doing without representing?" *Synthese* 101: 401–431.
Collins, A. and Quillian, M. 1972. "Experiments on semantic memory and language comprehension," in L. Gregg (ed.), *Cognition in Learning and Memory*. New York: Wiley, pp. 117–137.
Copeland, J. 1993. *Artificial Intelligence: A Philosophical Introduction*. Oxford: Blackwell.
1996. "What is computation?" *Synthese* 108: 335–359.
Crane, T. 2003. *The Mechanical Mind*. 2nd edn. London: Routledge.
Crowe, M. 2001. *Theories of the World from Antiquity to the Copernican Revolution: Revised Edition*. New York: Dover Publications.
Cummins, R. 1975. "Functional analysis," *Journal of Philosophy*, 72 (20): 741–756.
1983. *The Nature of Psychological Explanation*. Cambridge, MA: MIT Press.
1986. "Inexplicit information," in M. Brand and R. Harnish (eds.), *The Representation of Knowledge and Belief*. Tucson, AZ: University of Arizona Press, pp. 116–126.
1989. *Meaning and Mental Representation*. Cambridge, MA: MIT Press.

1991. "The role of representation in connectionist explanations of cognitive capacities," in W. Ramsey, S. Stich, and D. Rumelhart (eds.), *Philosophy and Connectionist Theory*. Hillsdale, NJ: Lawrence Erlbaum, pp. 91–114.

1996. *Representations, Targets and Attitudes*. Cambridge, MA: MIT Press.

de Charms, R. C. and Zador, A. 2000. "Neural representation and the cortical code," *The Annual Review of Neuroscience* 23: 613–647.

Delaney, C. F. 1993. *Science, Knowledge, and Mind: A Study in the Philosophy of C.S. Peirce*. Notre Dame: Notre Dame University Press.

Dennett, D. 1978. *Brainstorms*. Cambridge, MA: MIT Press.

1982. "Styles of mental representation," *Proceedings of the Aristotelian Society*, n.s. 83: 213–226. Reprinted in Dennett, 1987.

1987. *The Intentional Stance*. Cambridge, MA: MIT Press.

1990. "The myth of original intentionality," in K. A. Mohyeldin Said, W. H. Newton-Smith, R. Viale, and K. V. Wilkes (eds.), *Modelling the Mind*. Oxford: Clarendon Press, pp. 43–62.

1991a. "Two contrasts: folk craft versus folk science, and belief versus opinion," in J. D. Greenwood (ed.), *The Future of Folk Psychology: Intentionality and Cognitive Science*. Cambridge: Cambridge University Press, pp. 135–148.

1991b. *Consciousness Explained*. London: Penguin.

Dennett, D. and Haugeland, J. 1987. "Intentionality," in R. L. Gregory (ed.), *The Oxford Companion to the Mind*. Oxford: Oxford University Press, pp. 383–386.

Donahue, W. H. 1981. *The Dissolution of the Celestial Spheres*. Manchester, NH: Ayer Co Publishing.

Dretske, F. 1988. *Explaining Behavior*. Cambridge, MA: MIT Press.

Elliffe, M. 1999. "Performance measurement based on usable information," in R. Baddeley, P. Hancock, and P. Foldiak (eds.), *Information Theory and the Brain*. Cambridge: Cambridge University Press.

Field, H. 1978. "Mental representation," *Erkenntnis* 13: 9–61.

Fodor, J. 1968. *Psychological Explanations: An Introduction to the Philosophy of Psychology*. New York: Random House.

1980. "Methodological solipsism considered as a research strategy in cognitive science," *Behavioral and Brain Sciences* 3 (1): 63–73. Reprinted in Fodor, 1981.

1981. *RePresentations*. Cambridge, MA: MIT Press.

1985. "Fodor's guide to mental representation," *Mind* 94: 76–100.

1987. *Psychosemantics*. Cambridge, MA: MIT Press.

1990. *A Theory of Content and Other Essays*. Cambridge, MA: MIT Press.

1992. "The big idea: can there be a science of mind?" *Times Literary Supplement* July 3, pp. 5–7.

Fodor, J and Pylyshyn, Z. 1988. "Connectionism and cognitive architecture: a critical analysis," *Cognition* 28: 3–71.

Forster, M. and Saidel, E. 1994. "Connectionism and the fate of folk psychology: a reply to Ramsey, Stich and Garon," *Philosophical Psychology* 7 (4): 437–452.

Frankish, K. 2004. *Mind and Supermind*. Cambridge: Cambridge University Press.

Freeman, W. and Skarda, C. 1990. "Representations: who needs them?" in J. McGaugh, N. Weinberger, and G. Lynch (eds.), *Brain Organization and Memory: Cells, Systems and Circuits*. Oxford: Oxford University Press, pp. 375–380.

Gallistel, C. R. 1998. "Symbolic processes in the brain: the case of insect navigation," in D. Scarborough and S. Sternberg (eds.), *Methods, Models and Conceptual Issues: Vol. 4, of An Invitation to Cognitive Science*, 2nd edn. Cambridge, MA: MIT Press, pp. 1–51.

Goldman, A. 1992. "In defense of simulation theory," *Mind and Language* 7: 104–119.

Gopnik, A. and Wellman, H. 1992. "Why the child's theory of mind really is a theory," *Mind and Language* 7: 145–171.

Gordon, R. 1986. "Folk psychology as simulation," *Mind and Language* 1: 158–171.

Gorman, R. and Sejnowski, T. 1988. "Analysis of the hidden units in a layered network trained to classify sonar targets," *Neural Networks* 1: 75–89.

Grice, P. 1957. "Meaning," *Philosophical Review* 66: 377–388.

Griffiths, P. E. 2001. "Genetic information: a metaphor in search of a theory?" *Philosophy of Science* 68: 394–412.

Grush, R. 1997. "The architecture of representation," *Philosophical Psychology* 10 (1): 5–25.

2004. "The emulation theory of representation: motor control, imagery, and perception," *Behavioral and Brain Sciences* 27 (3): 377–396.

Haldane, J. 1993. "Understanding folk," in S. Christensen and D. Turner (eds.), *Folk Psychology and the Philosophy of Mind*. Hillsdale, NJ: Lawrence Erlbaum Associates, pp. 263–287.

Harnish, R. 2002. *Minds, Brains, and Computers*. Oxford: Blackwell Publishing.

Haugeland, J. 1978. "The nature and plausibility of cognitivism," *Behavioral and Brain Sciences* 2: 215–260. Reprinted in Haugeland, 1981.

1981. *Mind Design*. Cambridge, MA: MIT Press.

1985. *Artificial Intelligence: The Very Idea*. Cambridge, MA: MIT Press.

1991. "Representational genera," in W. Ramsey, S. Stich, and D. Rumelhart (eds.), *Philosophy and Connectionist Theory*. Hillsdale, NJ: Lawrence Erlbaum, pp. 61–89.

1997. *Mind Design II*. Cambridge, MA: MIT Press.

Haybron, D. 2000. "The causal and explanatory role of information stored in connectionist networks," *Minds and Machines* 10 (3): 361–380.

Heil, J. 1991. "Being indiscrete" in J. D. Greenwood (ed.), *The Future of Folk Psychology: Intentionality and Cognitive Science*. Cambridge: Cambridge University Press, pp. 120–134.

Homme, L. E. 1965. "Control of coverants, the operants of the mind. Perspectives in psychology, 24", *The Psychological Record* 15: 501–511.

Horgan, T. 1989. "Mental quasation," in J. Tomberlin (ed.), *Philosophical Perspectives* 3: 47–76.

1994. "Computation and mental representation," in S. Stich and T. Warfield (eds.), *Mental Representation: A Reader*. Oxford: Blackwell, pp. 302–311.

Hubel, D. and Wiesel, T. 1962. "Receptive fields, binocular interaction, and functional architecture in the cat's visual cortex," *Journal of Physiology* 160: 106–154.

1968. "Receptive fields and functional architecture of monkey striate cortex," *Journal of Physiology* 195: 215–243.

Hull, C. L. 1930. "Knowledge and purpose as habit mechanisms," *Psychological Review* 37: 511–525.

1931. "Goal attraction and directing ideas conceived as habit phenomena," *Psychological Review* 38: 487–506.

1932. "The goal gradient hypothesis and maze learning," *Psychological Review* 39: 25–43.

Jaynes, J. 1976. *The Origin of Consciousness in the Breakdown of the Bicameral Mind.* Wilmington, MA: Houghton Mifflin.

Johnson-Laird, P. 1983. *Mental Models: Towards a Cognitive Science of Language, Inference, and Consciousness.* Cambridge, MA: Harvard University Press.

Kendler, H. H. 1971. "Environmental and cognitive control of behavior," *American Psychologist* 26 (11): 962–973.

Kim, J. 1998. *Mind in a Physical World.* Cambridge, MA: MIT Press.

Kirsh, D. 1990. "When is information explicitly represented?" *The Vancouver Studies in Cognitive Science*, vol. 1: 340–365.

Kuhn, T. 1962. *The Structure of Scientific Revolutions.* Chicago: University of Chicago Press.

Lashley, K. 1960. "In search of the engram," in F. Beach, D. Hebb, C. Morgan, and H. Nissen (eds.), *The Neuropsychology of Lashley, Selected Papers.* New York: McGraw-Hill, pp. 478–505.

Leibniz, G. W. 1956. *Philosophical Papers and Letters.* L. Loemkar (ed.). Chicago: University of Chicago Press.

Lettvin, J., Maturana, H., McCulloch, W., and Pitts, W. 1959. "What the frog's eye tells the frog's brain," *Proceedings of the Institute of Radio Engineers* 47: 1940–1951.

Lloyd, D. 1995. "Consciousness: a connectionist manifesto," *Minds and Machines* 5 (2): 161–185.

Lycan, W. 1986. "Tacit belief," in R. J. Bogdan (ed.), *Belief.* Oxford: Clarendon Press, pp. 61–82.

Marr, D. 1982. *Vision.* San Francisco: W. H. Freeman.

Maynard-Smith, J. 2000. "The concept of information in biology," *Philosophy of Science* 67 (2): 177–194.

McNaughton, B. L. 1989. "Neural mechanisms for spatial computation and information storage," in L. A. Nadel, P. Cooper, P. Culicover, and R. Harnish (eds.), *Neural Connections and Mental Computations.* Cambridge, MA: MIT Press, pp. 285–349.

Melden, A. I. 1961. *Free Action.* New York: Humanities Press.

Millikan, R. 1984. *Language, Thought and Other Biological Categories.* Cambridge, MA: MIT Press.

1993. *White Queen Psychology and Other Essays for Alice.* Cambridge, MA: MIT Press.

1996. "Pushmi-pullyu representations," in J. Tomberlin (ed.), *Philosophical Perspectives IX: AI, Connectionism, and Philosophical Psychology*. Atascadero, CA: Ridgeview Publishing, pp. 185–200.

Mumford, S. 1998. *Dispositions*. Oxford: Clarendon Press.

Newell, A. 1980. "Physical symbol systems," *Cognitive Science* 4: 135–183.

1990. *Unified Theories of Cognition*. Cambridge, MA: Harvard University Press.

Newell, A. and Simon, H. 1976. "Computer science as empirical inquiry," *Communications of the ACM* 19 (3): 113–126. Reprinted in Haugeland, 1981.

Nisbett, R. and Wilson, T. 1977. "Telling more than we can know: verbal reports on mental processes," *Psychological Review* 84: 231–259.

O' Brien, G. and Opie, J. 1999. "A connectionist theory of phenomenal experience," *Behavioral and Brain Sciences* 22: 127–148.

Osgood, C. E. 1956. "Behavior theory and the social sciences," *Behavioral Sciences* 1: 167–185.

O' Reilly, R. C. and Munakata, Y. 2000. *Computational Explorations in Cognitive Neuroscience*. Cambridge, MA: MIT Press.

Palmer, S. 1978. "Fundamental aspects of cognitive representation," in E. Rosch and E. Lloyd (eds.), *Cognition and Categorization*. Hillsdale, NJ: Lawrence Erlbaum, pp. 259–303.

Papideau, D. 1984. "Representation and explanation," *Philosophy of Science* 51 (4): 550–572.

Peirce, C. S. 1931–58. *The Collected Papers of C.S. Peirce, vols. 1–8*. A. Burks, C. Hartshorne, and P. Weiss (eds.). Cambridge, MA: Harvard.

Pylyshyn, Z. 1984. *Computation and Cognition*. Cambridge, MA: MIT Press.

Ramsey, W. 1991. "Where does the self-refutation objection take us?" *Inquiry* 33: 453–465.

1995. "Rethinking distributed representation," *Acta Analytica* 14: 9–25.

1996. "Investigating commonsense psychology," *Communication and Cognition* 29 (1): 91–120.

1997. "Do connectionist representations earn their explanatory keep?" *Mind and Language* 12: 34–66.

2003. "Eliminative materialism," *The Stanford Encyclopedia of Philosophy*. E. N. Zalta (ed.), http://plato.stanford.edu.archives/fall2003/entries/materialism-eliminativism/.

Ramsey, W., Stich, S., and Garon, J. 1990. "Connectionism, eliminativism and the future of folk psychology," *Philosophical Perspectives* 4. Atascadero, CA: Ridgeview Publishing: 499–533. Reprinted in Ramsey, Stich, and Rumelhart, 1991.

Ramsey, W., Stich, S., and Rumelhart, D. 1991. *Philosophy and Connectionist Theory*. Hillsdale, NJ: Lawrence Erlbaum.

Reike, F., Warland, D., de Ruyter van Steveninck, R., and Bialek, W. 1997. *Spikes*. Cambridge, MA: MIT Press.

Robinson, W. 1992. *Computers, Minds and Robots*. Philadelphia: Temple University Press.

Rogers, T. and McClelland, J. 2004. *Semantic Cognition: A Parallel Distributed Processing Approach.* Cambridge, MA: MIT Press.

Roitblat, H. L. 1982. "The meaning of representation in animal memory," *Behavioral and Brain Sciences* 5, 353–406.

Rosch, E. and Mervis, C. B. 1975. "Family resemblances: studies in the internal structure of categories," *Cognitive Psychology* 7: 573–605.

Rumelhart, D. 1990. "Brain style computation: learning and generalization," in S. Zornetzer, J. Davis, and C. Lau (eds.), *An Introduction to Neural and Electronic Networks.* San Diego, CA: Academic Press, pp. 405–420.

Rumelhart, D. and McClelland, J. 1986a. *Parallel Distributed Processing, Vol. 1.* Cambridge, MA: MIT Press.

1986b. *Parallel Distributed Processing, Vol. 2.* Cambridge, MA: MIT Press.

Ryder, D. 2004. "SINBAD neurosemantics: a theory of mental representation," *Mind and Language* 19 (2): 211–240.

Ryle, G. 1949. *The Concept of Mind.* Chicago: University of Chicago Press.

Searle, J. 1980. "Minds, brains and programs," *Behavioral and Brain Sciences* 3: 417–424. Reprinted in Haugeland, 1997.

1983. *Intentionality: An Essay in the Philosophy of Mind.* Cambridge: Cambridge University Press.

1984. *Minds, Brains and Science.* Cambridge, MA: Harvard University Press.

1990. "Is the brain a digital computer?" *Proceedings and Addresses of the American Philosophical Association* 64: 21–37.

1991. *The Rediscovery of the Mind.* Cambridge, MA: MIT Press.

Sejnowski, T. J. and Rosenberg, C. R. 1987. "Parallel networks that learn to pronounce English text," *Complex Systems* 1: 145–168.

Shank, R. and Abelson, R. 1977. *Scripts, Plans, Goals and Understanding.* Hillsdale, NJ: Lawrence Erlbaum.

Shannon, C. and Weaver, W. 1949. *The Mathematical Theory of Communication.* Chicago: University of Illinois Press.

Shapiro, M. and Eichenbaum, H. 1997. "Learning and memory: computational principles and neural mechanisms," in M. D. Rugg (ed.), *Cognitive Neuroscience.* Cambridge, MA: MIT Press.

Singer, P. 1972. "Famine, affluence and morality," *Philosophy and Public Affairs* 1: 229–243.

Skinner, B. F. 1976. *About Behaviorism.* New York: Vintage Books.

Smith, E. and Medin, D. 1981. *Categories and Concepts.* Cambridge, MA: Harvard.

Smolensky, P. 1988. "On the proper treatment of connectionism," *Behavioral and Brain Sciences* 11: 1–23.

1991. "Connectionism, constituency, and the language of thought," in B. Loewer and G. Rey (eds.), *Meaning in Mind: Fodor and his Critics.* Oxford: Basil Blackwell, pp. 201–227.

Snyder, L., Batista, A., and Anderson, R. A. 1997. "Coding of intention in the posterior parietal cortex," *Nature* 386: 167–170.

Stalnaker, R. 1984. *Inquiry.* Cambridge, MA: MIT Press.

Sterelney, K. 2003. *Thought in a Hostile World: The Evolution of Human Cognition.* Malden, MA: Blackwell Publishing.

Stich, S. 1983. *From Folk Psychology to Cognitive Science: The Case Against Belief.* Cambridge, MA: MIT Press.

1992. "What is a theory of mental representation?" *Mind* 101: 243–261. Reprinted in S. Stich and T. Warfield (eds.), 1994.

1996. *Deconstructing the Mind.* Oxford: Oxford University Press.

Stich, S. and Nichols, S. 1993. "Folk psychology: simulation or tacit theory?" *Mind and Language* 7: 35–71.

Stich, S. and Warfield, T. (eds.) 1994. *Mental Representation: A Reader.* Oxford: Blackwell.

Swoyer, C. 1991. "Structural representation and surrogative reasoning," *Synthese* 87: 449–508.

Thelan, E. and Smith, L. 1994. *A Dynamic Systems Approach to the Development of Cognition and Action.* Cambridge, MA: MIT Press.

Unger, P. 1996. *Living High and Letting Die.* Oxford: Oxford University Press.

van Gelder, T. 1991. "What is the 'D' in 'PDP'?, a survey of the concept of distribution," in W. Ramsey, S. Stich, and D. Rumelhart (eds.), *Philosophy and Connectionist Theory.* Hillsdale, NJ: Lawrence Erlbaum, pp. 33–59.

1995. "What might cognition be, if not computation?" *The Journal of Philosophy* 91: 345–381.

Von Eckardt, B. 1993. *What is Cognitive Science?* Cambridge, MA: MIT Press.

Warfield, T. 1998. "Commentary on John Searle," *Notre Dame Perspectives Lecture.*

Watson, J. B. 1930. *Behaviorism* (revised edn.). Chicago: University of Chicago Press.

White, P. A. 1988. "Knowing more than we can tell: 'introspective access' and causal report accuracy ten years later," *British Journal of Psychology* 79 (1): 13–45.

Winograd, T. 1972. *Understanding Natural Language.* New York: Academic Press.

Index